Julia Verena Hartig

Functional analysis of Loqs and R2D2 in small RNA silencing pathways

AF011297

Julia Verena Hartig

Functional analysis of Loqs and R2D2 in small RNA silencing pathways

The role of small double-stranded RNA binding domain proteins in the network of silencing pathways in Drosophila

Südwestdeutscher Verlag für Hochschulschriften

Impressum/Imprint (nur für Deutschland/ only for Germany)
Bibliografische Information der Deutschen Nationalbibliothek: Die Deutsche Nationalbibliothek verzeichnet diese Publikation in der Deutschen Nationalbibliografie; detaillierte bibliografische Daten sind im Internet über http://dnb.d-nb.de abrufbar.

Alle in diesem Buch genannten Marken und Produktnamen unterliegen warenzeichen-, marken- oder patentrechtlichem Schutz bzw. sind Warenzeichen oder eingetragene Warenzeichen der jeweiligen Inhaber. Die Wiedergabe von Marken, Produktnamen, Gebrauchsnamen, Handelsnamen, Warenbezeichnungen u.s.w. in diesem Werk berechtigt auch ohne besondere Kennzeichnung nicht zu der Annahme, dass solche Namen im Sinne der Warenzeichen- und Markenschutzgesetzgebung als frei zu betrachten wären und daher von jedermann benutzt werden dürften.

Verlag: Südwestdeutscher Verlag für Hochschulschriften GmbH & Co. KG
Dudweiler Landstr. 99, 66123 Saarbrücken, Deutschland
Telefon +49 681 37 20 271-1, Telefax +49 681 37 20 271-0
Email: info@svh-verlag.de
Zugl.: München, LMU, Diss., 2010

Herstellung in Deutschland:
Schaltungsdienst Lange o.H.G., Berlin
Books on Demand GmbH, Norderstedt
Reha GmbH, Saarbrücken
Amazon Distribution GmbH, Leipzig
ISBN: 978-3-8381-2476-6

Imprint (only for USA, GB)
Bibliographic information published by the Deutsche Nationalbibliothek: The Deutsche Nationalbibliothek lists this publication in the Deutsche Nationalbibliografie; detailed bibliographic data are available in the Internet at http://dnb.d-nb.de.

Any brand names and product names mentioned in this book are subject to trademark, brand or patent protection and are trademarks or registered trademarks of their respective holders. The use of brand names, product names, common names, trade names, product descriptions etc. even without a particular marking in this works is in no way to be construed to mean that such names may be regarded as unrestricted in respect of trademark and brand protection legislation and could thus be used by anyone.

Publisher: Südwestdeutscher Verlag für Hochschulschriften GmbH & Co. KG
Dudweiler Landstr. 99, 66123 Saarbrücken, Germany
Phone +49 681 37 20 271-1, Fax +49 681 37 20 271-0
Email: info@svh-verlag.de

Printed in the U.S.A.
Printed in the U.K. by (see last page)
ISBN: 978-3-8381-2476-6

Copyright © 2011 by the author and Südwestdeutscher Verlag für Hochschulschriften GmbH & Co. KG and licensors
All rights reserved. Saarbrücken 2011

Contents

1 **SUMMARY** 5
2 **INTRODUCTION** 7
    2.1 CLASSES OF SMALL RNAS 7
        2.1.1 *Micro RNAs (miRNAs)* 7
        2.1.2 *Exogenous siRNAs (exo-siRNAs)* 9
        2.1.3 *Endogenous siRNAs (endo-siRNAs)* 10
        2.1.4 *Piwi-interacting RNAs (piRNAs)* 10
    2.2 THE SIGNIFICANCE OF SMALL RNA SILENCING 11
        2.2.1 *Roles of small RNA silencing pathways* 11
        2.2.2 *Transposable elements* 12
    2.3 THE PROBLEM OF PATHWAY SPECIFICITY 13
        2.3.1 *dsRBPs as specificity factors for small RNA sorting* 13
            2.3.1.1 Properties of double-stranded RNA binding domains 13
            2.3.1.2 R2D2 14
            2.3.1.3 Loquacious 14
        2.3.2 *Sorting and specificity problems* 14
            2.3.2.1 Sorting between miRNAs and exo-siRNAs 14
            2.3.2.2 Sorting between miRNAs and endo-siRNAs 15
            2.3.2.3 Sorting between endo-siRNAs and exo-siRNAs 15
    2.4 DROSOPHILA GENETICS 15
        2.4.1 *GFP-based cell culture reporter systems* 15
        2.4.2 *Transgenic flies* 17
            2.4.2.1 Embryo injection and transgenic fly lines 17
            2.4.2.2 The UAS/Gal4 expression system 17
    2.5 SMALL RNA SILENCING SYSTEMS IN OTHER ORGANISMS 19
3 **SPECIFIC AIMS OF THIS THESIS** 21
4 **MATERIALS AND METHODS** 22
    4.1 MATERIALS 22
        4.1.1 *Laboratory hardware* 22
        4.1.2 *Analysis software* 22
        4.1.3 *Laboratory chemicals* 23
        4.1.4 *Radiochemicals* 24
        4.1.5 *Enzymes* 25
            4.1.5.1 General enzymes 25
            4.1.5.2 Polymerases 25
            4.1.5.3 Restriction enzymes 25
        4.1.6 *Kits* 25
        4.1.7 *Other materials* 26
        4.1.8 *Plasmids in laboratory stock* 27
        4.1.9 *Cells* 28
            4.1.9.1 Bacterial stocks 28
            4.1.9.2 Cell lines 29
        4.1.10 *Fly stocks* 29
        4.1.11 *PCR primers* 30
            4.1.11.1 Cloning 30
            4.1.11.2 Sequencing 34
            4.1.11.3 qPCR 34

	4.1.11.4	Test-PCR	35
	4.1.11.5	RACE	35
	4.1.11.6	Mapping P-element insertions in transgenic flies	35
	4.1.11.7	Sequencing primer for pUASP-trangenic flies	36
4.1.12		*Media*	*37*
	4.1.12.1	Bacterial stocks	37
	4.1.12.2	Cell culture	37
4.1.13		*Fly food*	*38*
4.1.14		*Antibodies*	*38*
	4.1.14.1	Primary antibodies	38
	4.1.14.2	Secondary antibodies	39
4.1.15		*Stock solutions and commonly used buffers*	*39*
4.2		**METHODS**	**43**
4.2.1		*Molecular cloning*	*43*
	4.2.1.1	Primer design for cloning of dsRBDs	43
	4.2.1.2	Amplification of DNA sequences by Polymerase Chain Reaction (PCR)	43
	4.2.1.3	Agarose gel electrophoresis	44
	4.2.1.4	Specific digestion of DNA by restriction endonucleases	44
	4.2.1.5	Ligation of vector with insert DNA	44
	4.2.1.6	Bacterial transformation	45
	4.2.1.7	Test for correct transformants by colony-PCR	45
	4.2.1.8	Preparation of plasmid DNA	45
	4.2.1.9	DNA sequencing	45
	4.2.1.10	3´-RACE PCR analysis of the loqs-RD variant	45
4.2.2		*Methods of Drosophila S2 cell culture*	*46*
	4.2.2.1	Maintenance	46
	4.2.2.2	Depletion of individual genes by RNAi in cell culture	46
	4.2.2.3	Selection of clonal cell lines	48
	4.2.2.4	Storage of cells in liquid nitrogen	48
4.2.3		*Protein analysis*	*48*
	4.2.3.1	Protein extraction	48
	4.2.3.2	Co-immunoprecipitation	49
	4.2.3.3	Immunoblotting for detection of proteins	49
	4.2.3.4	α-Loqs-PD-specific antibody production	50
	4.2.3.5	Dot blot	50
4.2.4		*RNA analysis*	*50*
	4.2.4.1	RNA extraction	50
	4.2.4.2	Northern Blotting	50
	4.2.4.3	Analysis of mRNA levels by Polymerase Chain Reaction	51
4.2.5		*Drosophila melanogaster methods*	*53*
	4.2.5.1	Maintenance and handling	53
	4.2.5.2	Transgenic flies	54
4.2.6		*Recombinant expression and purification of GST- or His$_6$-tagged Loqs isoforms*	*55*
	4.2.6.1	Recombinant expression	55
	4.2.6.2	Affinity purification of recombinant proteins	56

5 RESULTS 57

5.1	A NOVEL ISOFORM OF *LOQS*	57
5.2	ISOFORM-SPECIFIC KNOCK-DOWN	59
5.3	A CELL CULTURE REPORTER SYSTEM FOR ENDO-SIRNA SILENCING ACTIVITY	63
5.4	LOQS-PD IS ESSENTIAL FOR SILENCING HIGH-COPY TRANSGENES	66
5.5	LOQS-PD IS ESSENTIAL FOR BIOGENESIS OF HAIRPIN-DERIVED ENDO-SIRNAS	68
5.6	LOQS-PD-SPECIFIC ANTIBODY	70

5.7	Loqs-PD interacts with Dcr-2 in cell culture and flies	71
5.8	The PD-specific C-terminus is sufficient for Dcr-2 binding and essential for endo-siRNA function	73
5.9	Loqs-PD interacts with the N-terminal helicase domain of Dcr-2 during endo-siRNA biogenesis	76
5.10	R2D2 acts as an antagonist of Loqs-PD in endo-siRNA silencing	81
5.11	The role of Loqs-PD in exo-siRNA silencing	83
5.12	Multimerization and competition of Loqs isoforms for Dcr-2 binding	86
5.13	Transcriptional vs. post-transcriptional gene silencing	89
5.14	Comparison between two clonal cell lines: 63-6 and 63N1	90
5.15	Loqs-PD associates with Dcr-2 *in vivo*	92
5.16	Loqs-PD is essential for endo-siRNA biogenesis *in vivo*	94
5.17	Future perspective	97
6	**DISCUSSION**	**102**
6.1	Distinct Loqs isoforms separate the biogenesis routes for endo-siRNAs and miRNAs	102
6.2	Antagonism of small RNA biogenesis pathways	103
6.3	Competition between Loqs isoforms	106
6.4	A conserved interaction scheme between Dicer and dsRBD proteins	108
6.5	Affinity *versus* abundance: Stability of the siRNA system and a model for siRNA precursor recognition	110
6.6	Comparison between two reporter cell lines: 63N1 and 63-6	111
6.7	Transcriptional *versus* post-transcriptional gene silencing	112
6.8	Is there an RdRP-like activity in Drosophila?	112
7	**ABBREVIATIONS**	**114**
8	**APPENDIX**	**118**
9	**ACKNOWLEDGEMENTS/DANKSAGUNG**	**128**
10	**REFERENCES**	**130**

1 Summary

Small non-coding RNA-dependent gene silencing is a highly conserved mechanism found in fungi, plants and animals. *Drosophila melanogaster* is one of the best studied model-organisms for small RNA-dependent silencing. A variety of small RNA classes are involved in regulating a wide array of cellular processes ranging from development to cancer. Micro RNAs (miRNAs) are genomically encoded and repress expression of endogenous genes. Exogenous small interfering RNAs (exo-siRNAs) serve to defend cells from viral infections and are widely used in artificial gene silencing by RNA interference (RNAi). Piwi-interacting RNAs (piRNAs) act to suppress transposable elements in the germ-line. Most recently, endogenous small interfering RNAs (endo-siRNAs) were discovered to silence transposons in somatic cells.

Leaving the special biogenesis of piRNAs aside, there are certain elements that are common to miRNAs, exo-siRNAs and endo-siRNAs. All three have to be processed from longer RNA precursors by pairs of an RNAseIII class endonuclease together with a double-stranded RNA binding-domain protein (dsRBP). Processed short double-stranded precursors are loaded into effector complexes where they become single-stranded and target the silencing machinery to a complementary mRNA. Since dsRBPs can interact with both the RNA substrate and the RNAseIII enzyme they are considered to be specificity factors that contribute to faithful sorting of small RNAs into their respective effectors.

siRNA mediated silencing in *Drosophila* can be subdivided into exo- and endo-siRNA dependent pathways. In both cases 21 nt siRNAs are excised from stretches of long double-stranded precursors by the RNaseIII endonuclease Dcr-2 and loaded into the Argonaute protein Ago2. The pathways are set apart by the double-stranded RNA binding domain proteins they require: R2D2 is needed to load exo-siRNAs into Ago2 while this thesis analyzes a novel isoform of Loquacious, Loqs-PD, which is required for endo-siRNA silencing. Its properties are distinct from the described Loqs-PB isoform which acts in miRNA biogenesis. Loqs-PD arises from an alternatively poly-adenylated variant of the gene and has a stretch of 22 isoform-specific amino acids at its C-terminus. The C-terminal peptide is sufficient to interact with the DExH/D-helicase domain of *Drosophila* Dcr-2. This association can be found *in vitro* and *in vivo* and is essential for endo-siRNA biogenesis and silencing of artificially introduced transgenes. In Schneider 2 *Drosophila* cell culture cells (S2 cells), R2D2 acts as an antagonist of Loqs-PD in endo-siRNA silencing and is not required for loading of endo-siRNAs into Ago2. Other isoforms of Loqs (Loqs-PA and Loqs-PB) preferentially bind to Dcr-1 but can compete with Loqs-PD for binding to Dcr-2, albeit with lower affinity. Thus my findings illustrate how specificity between the small RNA-dependent silencing pathways is achieved.

2 Introduction

2.1 Classes of small RNAs

Since the discovery of small RNA silencing little more than a decade ago, an increasingly complicated variety of classes and pathways has been described. The most important ones in animals include micro RNAs (miRNAs), small-interfering RNAs (siRNAs) and Piwi-interacting RNAs (piRNAs). miRNAs and siRNAs were the first small RNA species to be discovered (Lee et al., 1993; Dalmay et al., 2000). Most of the small RNA silencing pathways are highly conserved and can be found in all eukaryotic phyla from the yeast *S. pombe* to plants and animals, but not in bacteria or archea (reviewed in Ghildiyal et al., 2009). Table 1 shows an overview of small RNA dependent silencing pathways in the fruit fly *Drosophila melanogaster*, one of the best studied model-systems for small RNA silencing.

2.1.1 Micro RNAs (miRNAs)

As depicted in Figure 1A, miRNAs are encoded in the genome (Bartel, 2004a) and – in most cases – transcribed by RNA Polymerase (Pol II) (Lee et al., 2004). The transcripts, called pri-miRNAs, form hairpins with imperfect complementarity (Lee et al., 2002). Two pairs of an RNAseIII enzyme and a double-stranded RNA binding domain protein (dsRBP) then process the primary miRNA to form a double-stranded precursor: Drosha and Pasha perform the first cleavage in the nucleus, Dicer-1 (Dcr-1) and its dsRBP partner Loquacious (Loqs) remove the hairpin loop in the cytoplasm (Forstemann et al., 2005; Jiang et al., 2005; Saito et al., 2005; Park et al., 2007). The double-stranded precursor is then loaded into an effector complex, termed RNA-induced silencing complex (RISC). In *Drosophila* most miRNAs are loaded into the effector endonuclease Argonaute 1 (Ago1) (Lee et al., 2004; Okamura et al., 2004). Depending on thermodynamic characteristics of the precursor, one strand, the miRNA*, is then expelled from the complex. After that complementary base-pairing with an mRNA can silence gene expression by inhibiting translation or favoring degradation of the message (Okamura et al., 2004).

Table 1: Overview of small RNA silencing pathways in *Drosophila melanogaster*

small RNA class	origin	precursor structure	mature length	processing machinery	RISC loading complex	Argonaute protein	mode of silencing
miRNA	endogenous	~70 nt hairpin	~22 nt	Dcr-1/Loqs	?	Ago1	translational repression/mRNA degradation
exo-siRNA	external; viruses, RNAi	long dsRNA	~21 nt	Dcr-2 (R2D2)	Dcr-2/R2D2	Ago2	mRNA and viral RNA cleavage ("slicing")
endo-siRNA	endogenous	long hairpin/long dsRNA from convergent transcription	~21 nt	Dcr-2/Loqs	?	Ago2	transposon silencing in somatic cells
piRNA	endogenous; clustered in piRNA master loci	single-stranded transposon transcript	24-30 nt	Piwi-subfamily Argonaute proteins (Aub, Piwi; Ago3); ping-pong mechanism		Piwi-subfamily Argonaute protein (Aub, Piwi; Ago3); ping-pong mechanism	transposon silencing in the germline

2.1.2 Exogenous siRNAs (exo-siRNAs)

siRNAs are derived from long double-stranded precursors (Figure 1B, left). These are either introduced into the cell with the purpose of inducing RNAi or appear during replication of certain RNA viruses (reviewed in Golden et al., 2008). In a similar biogenesis pathway as before cleavage by an RNAseIII enzyme, Dcr-2, with the help of a dsRBP, R2D2, yields a double-stranded precursor (Liu et al., 2003). Due to their origins from long double-stranded RNA siRNA precursors are perfectly complementary. Therefore they are preferentially loaded into a RISC featuring the more efficient endonuclease Ago2 (Liu et al., 2003; Pham et al., 2004). Unlike canonical miRNAs, siRNAs thus silence their targets by cleavage of the corresponding message (Tuschl et al., 1999; Okamura et al., 2004).

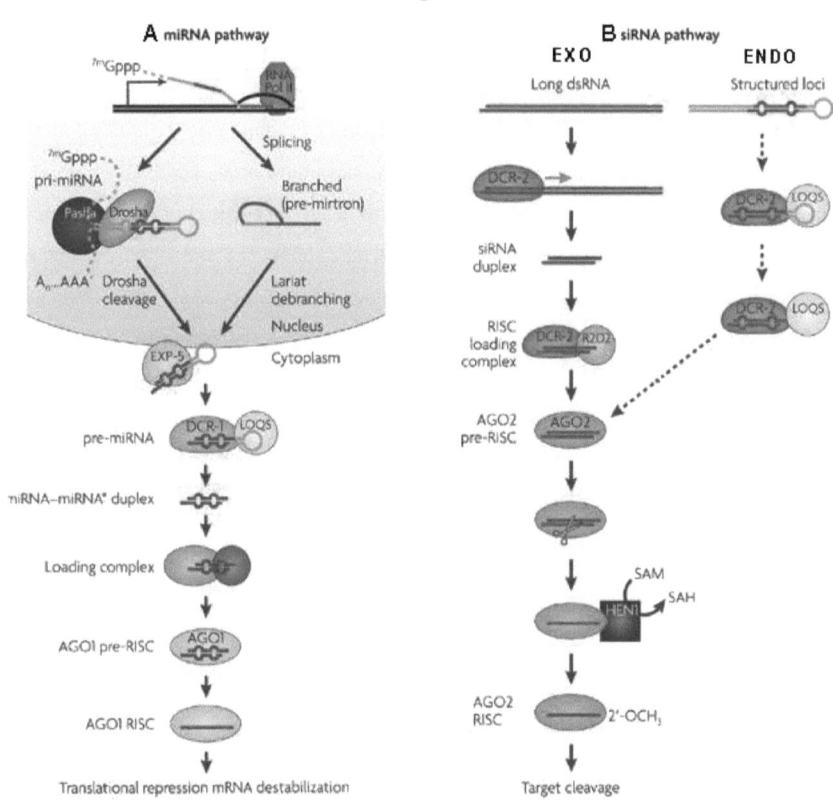

Figure 1: Exo-siRNA, endo-siRNA and miRNA silencing pathways in flies (legend continued on p. 13)

(legend Figure 1 continued)
The pathways differ in their substrates, biogeneses, effector proteins and modes of target regulation:
A) miRNAs are encoded in the genome and are transcribed to yield a primary miRNA (pri-miRNA) transcript, which is cleaved by Drosha to yield a short precursor miRNA (pre-miRNA). Alternatively, miRNAs can be present in introns (termed mirtrons) that are liberated following splicing to yield authentic pre-miRNAs. pre-miRNAs are exported from the nucleus to the cytoplasm, where they are further processed by DCR-1 to generate a duplex containing two strands, miRNA and miRNA*. Once loaded into AGO1, the miRNA strand guides translational repression of target RNAs.
B) *Left:* dsRNA precursors are processed by Dicer-2 (DCR-2) to generate siRNA duplexes containing guide and passenger strands. DCR-2 and the dsRNA-binding protein R2D2 (which together form the RISC-loading complex, RLC) load the duplex into Argonaute2 (AGO2).
Right: A subset of endogenous siRNAs (endo-siRNAs) exhibits dependence on dsRNA-binding protein Loquacious (LOQS), rather than on R2D2.
The passenger strand is later destroyed and the guide strand directs AGO2 to the target RNA.
Adapted from Ghildiyal et al., 2009

2.1.3 Endogenous siRNAs (endo-siRNAs)

Endo-siRNAs are derived from long double-stranded RNA precursors with endogenous origin, hence the term "endo" (Figure 1B, right). These are produced *in cis* from long hairpin structures with extensive stretches of complementarity (Okamura et al., 2008b), *in trans* from convergent transcription (Czech et al., 2008; Kawamura et al., 2008; Okamura et al., 2008a; Okamura et al., 2008c), or potentially by low levels of cryptic antisense transcription throughout the genome (reviewed in Berretta et al., 2009). The endo-siRNA pathway in *Drosophila* depends on Dicer-2 and Ago2 (Chung et al., 2008; Czech et al., 2008; Ghildiyal et al., 2008; Kawamura et al., 2008), the RNAseIII enzyme and the RISC factor usually associated with exo-siRNA biogenesis and function, hence the name "siRNA". Surprisingly, cell culture assays found involvement of Loqs instead of R2D2, the dsRBP acting with Dcr-2 in the exo-siRNA pathway (Czech et al., 2008; Kawamura et al., 2008; Okamura et al., 2008b).

2.1.4 Piwi-interacting RNAs (piRNAs)

Figure 2 shows that piRNA biogenesis in *Drosophila* is distinct from the other small RNA silencing pathways since it is Dicer-independent (Forstemann et al., 2005). Instead, Argonaute proteins of the germline-specific Piwi-subfamily (Le et al., 2007) take over the tasks of processing and target cleavage in a self-amplifying ping-pong mechanism (Brennecke et al., 2007; Gunawardane et al., 2007). Transposon anti-sense transcripts from hot-spots in the genome with large clusters of selfish genetic elements, so-called piRNA master loci, are cleaved at their 5´-ends by Ago3. After processing of the 3´-end by a yet unknown mechanism they can serve to prime the Piwi-class Argonaute proteins Aub and

Piwi to target transposon mRNAs. Cleavage products can then in turn serve as a new template for Ago3 (reviewed in Aravin et al., 2007; Hartig et al., 2007; Malone et al., 2009).

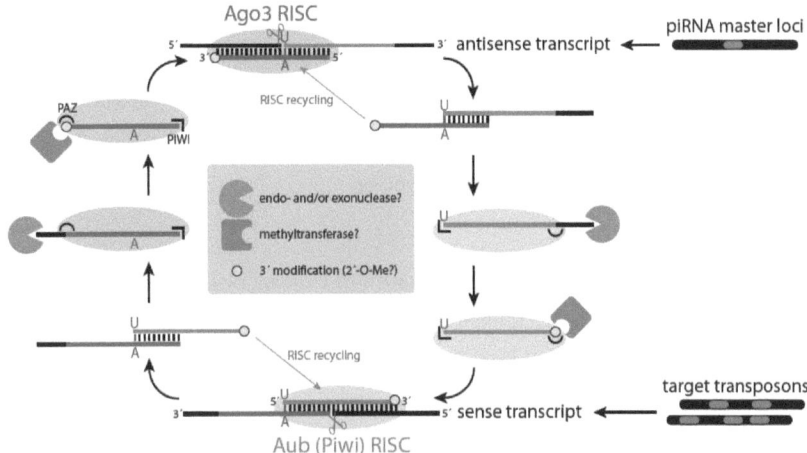

Figure 2: The ping-pong model for piRNA biogenesis
Sense transcripts from transposons – the kind that needs to be destroyed – are cleaved by Piwi or Aub RISCs loaded with a piRNA guide. The cleaved transcript is not merely degraded but used to program Ago3 RISC. This complex in turn cleaves the antisense transcripts that originate from the master control loci. Again, the cleaved RNA serves to program Piwi or Aub RISC. Thus, sense and antisense transcripts fuel an amplification cycle. While the 5'-ends of piRNAs are defined by RISC cleavage, the 3'-ends could be shortened by a 3' exonuclease until their length fits into the binding pocket between PAZ and PIWI domains. Presumably a 2'-O-Me is subsequently attached to the 3'-end by a yet unidentified methyltransferase.
Figure from Hartig et al., 2007

2.2 The significance of small RNA silencing

2.2.1 Roles of small RNA silencing pathways

Piwi-interacting RNAs counteract the mobilization of transposable elements in the germline (reviewed in Hartig et al., 2007; Malone et al., 2009). They are essential for genomic integrity of the offspring but require a maternally contributed pool of piRNAs for an efficient response (Blumenstiel et al., 2005; Czech et al., 2008). miRNAs regulate gene expression and are important in development and oncogenesis (reviewed in Ghildiyal et al., 2009; Kim et al., 2009). The evolutionary purpose of exo-siRNAs is defense against RNA viruses that involve the production of long dsRNA in their amplification cycle. The machinery can be exploited for RNA interference (RNAi), that is artificial gene silencing by introduction of double-stranded

RNA into the organism (reviewed in Ghildiyal et al., 2009; Kim et al., 2009). Endo-siRNAs suppress transposable elements in the soma and regulate expression of endogenous genes (Chung et al., 2008; Czech et al., 2008; Ghildiyal et al., 2008; Kawamura et al., 2008). Certain recurrent principles can be found among small silencing RNAs regarding types of precursors, modes of biogenesis and mechanisms of gene regulation. However, small RNA silencing pathways have distinct roles in maintaining optimal conditions for defense against external and internal threats. The focus of my thesis is a closer functional understanding of the endo-siRNA pathway and its role for transposon silencing.

2.2.2 Transposable elements

Transposons are nucleic acid "parasites" that are able to integrate into the genome of a host as well as mobilize and propagate themselves (Kazazian, 2004). This is highly detrimental to the genomic stability of an organism, especially if selfish genetic elements invade the germline. Genetic parasites can be found in all phyla but the prevalence of individual transposon classes varies between organisms. In the yeast *Saccharomyces cerevisiae*, only 3% of the genome are taken over by transposons (Kim et al., 1998), while in humans they constitute 50% and in corn even 80% of the genetic material (Flavell et al., 1974; Lander et al., 2001). Selfish genetic elements are characterized by their high copy number and their integration in tandem arrays in the genome. Their ability to excise themselves from one locus and integrate at another poses several immediate risks for the host (Deininger et al., 1999; Girard et al., 1999; Druker et al., 2004): First, integration in a promotor region or an open reading frame (ORF) can disrupt expression of a gene or deregulate its expression. Second, excision of a transposon leaves a DNA double-strand break that can lead to chromosome aberrations upon faulty repair. Last, tandem arrays of transposons cause recombination between non-homologous chromosomes during meiosis. Considering this wide array of disruptive effects it is surprising that only 1 in 600 germline mutations in humans is caused by transposons (Kazazian, 1999). This requires an effective mechanism to counter the threat of selfish genetic elements. In *Drosophila melanogaster* this role is taken over by piRNAs in the germline and endo-siRNAs in somatic cells.

2.3 The problem of pathway specificity

As indicated above miRNAs, exo-siRNAs and endo-siRNA depend on a mechanistically similar nucleolytic processing step in the cytoplasm carried out by a complex of Dicer and a double-stranded RNA binding domain protein (reviewed in Ghildiyal et al., 2009; Kim et al., 2009). The summary for the *Drosophila* system demonstrates the problem of specificity between the pathways: Dicer-1 together with the dsRBP Loquacious processes miRNAs (Forstemann et al., 2005; Jiang et al., 2005; Saito et al., 2005). In contrast, exo-siRNA precursors are processed by Dicer-2 and then loaded by the complex of Dcr-2 and R2D2 into Ago2 (Liu et al., 2003; Pham et al., 2004; Tomari et al., 2004). The endo-siRNA pathway combines factors from both miRNAs and exo-siRNAs: Dcr-2 interacts with Loqs, while an Ago2-RISC serves as the effector (Czech et al., 2008; Ghildiyal et al., 2008; Kawamura et al., 2008; Okamura et al., 2008b).

2.3.1 dsRBPs as specificity factors for small RNA sorting

2.3.1.1 Properties of double-stranded RNA binding domains

Double-stranded RNA binding domains (dsRBDs) are approximately 70 amino acid long domains with high evolutionary conservation. According to their similarity to a defined consensus they can be grouped into two types (Doyle et al., 2002): Type A dsRBDs are highly conserved in their entire length, while type B dsRBDs deviate from the consensus in their N-terminal region and bind dsRNA only poorly. dsRNA interaction is mediated via the sugar-phosphate backbone and is not sequence-specific (Ryter et al., 1998), however it can be sensitive to the structure and can recognize base-pairing mismatches, as can be seen by site-specific RNA-editing (Stefl et al., 2006). In addition to their RNA-binding activity, double-stranded RNA binding domains can mediate protein-protein interactions as well (Doyle et al., 2002). These properties make dsRBPs perfectly suited to act as specificity factors to funnel RNA precursors into the correct processing and effector complexes: They are able to recognize the double-stranded RNA precursor, possibly including the extent of complementary base pairing, and mediate protein-protein interaction with the corresponding Dicer enzyme for processing and subsequent loading into the appropriate RISC. R2D2 and Loquacious are the cytoplasmic dsRBDs in *Drosophila* that decide between miRNA, endo-siRNA or exo-siRNA fate.

2.3.1.2 R2D2

R2D2 contains two dsRBDs and a C-terminal part (later abbreviated R1R2RCterm). The latter mediates association with Dcr-2 (Liu et al., 2003; Ye et al., 2007). The complex of Dcr-2 and R2D2 does not have enhanced dsRNA processing activity; instead, it serves as the RISC loading complex (RLC) that loads double-stranded exo-siRNA precursors into Ago2 and interprets their thermodynamic asymmetry to select the incorporated strand of the mature siRNA (Pham et al., 2004; Tomari et al., 2004; Liu et al., 2006).

2.3.1.3 Loquacious

At the beginning of my work there were three known splice-variants of Loquacious (Forstemann et al., 2005). Loqs-PA and Loqs-PB both have three dsRBDs (L1L2L3) and can be found in adult flies as well as *Drosophila* Schneider cells (S2) (Forstemann et al., 2005; Miyoshi et al., 2009). The role of Loqs-PA is still largely uncharacterized, but Loqs-PB increases the efficiency of miRNA precursor processing by Dcr-1 (Forstemann et al., 2005; Jiang et al., 2005; Saito et al., 2005; Liu et al., 2007). Loqs-PC lacks the third dsRBD and instead carries a short peptide sequence at its C-terminus (abbreviated L1L2PCspec; 54 amino acids; Forstemann et al., 2005). The third dsRBD of Loqs is essential for Dcr-1 interaction, which is strongest for Loqs-PB but significant for Loqs-PA as well, while Loqs-PC does not form complexes with Dcr-1 (Forstemann et al., 2005; Ye et al., 2007).

2.3.2 Sorting and specificity problems

If we assume that dsRBPs are the decisive specificity factor for small RNA pathways the problem is immediately obvious: How can two dsRBPs, namely Loqs and R2D2, define three different pathways, namely miRNAs, endo-siRNAs and exo-siRNAs? Especially surprising was the promiscuity of Loqs which was reported to interact with both Dcr-1 and Dcr-2.

2.3.2.1 Sorting between miRNAs and exo-siRNAs

miRNA and exo-siRNA are processed by two completely distinct complexes: Typical miRNA-precursors with a certain degree of base-mismatches in the stem-structure are recognized by Dcr-1/Loqs, perfectly base-paired siRNA precursors by Dcr-2/R2D2. However, some highly complementary miRNAs are loaded into Ago2 RISCs (Förstemann, 2007; Seitz et al., 2008) and some siRNAs are more likely to enter Ago1 complexes (Lee et al., 2004; Forstemann et al., 2005). Thus it is not the processing step that determines the most suitable effector complex but the position and number of base-pairing mismatches in the double-stranded precursors (Förstemann, 2007; Tomari, 2007). Interestingly, only one isoform of Loqs, Loqs-PB, was necessary and sufficient for miRNA biogenesis (Forstemann et al., 2005; Park et al., 2007).

2.3.2.2 Sorting between miRNAs and endo-siRNAs

In order to act in both the miRNA- and the endo-siRNA pathway, Loqs would have to interact with both Dcr-1 and Dcr-2 as well as recognize completely distinct RNA substrates. A straightforward explanation to resolve this sorting dilemma would be that different Loqs isoforms are specific for each pathway. With Loqs-PB being the essential miRNA factor, this would leave Loqs-PA or -PC to act in endo-siRNA silencing.

2.3.2.3 Sorting between endo-siRNAs and exo-siRNAs

Endo-siRNA and exo-siRNA pathways are functionally distinct but mechanistically very similar: long perfectly complementary RNA precursors, albeit with different origins, dicing by the RNAseIII enzyme Dcr-2 and loading into Ago2 RISCs. Yet, it remains unclear how both R2D2 and a putative endo-siRNA-specific Loqs isoform bind Dcr-2 and how this binding influences the two pathways.

2.4 Drosophila genetics

Drosophila melanogaster is one of the most intensely studied model organisms with an array of established genetic tools. Additional advantages are the availability of immortalized cell lines and the possibility to produce and collect large amounts of material for biochemical analyses.

2.4.1 GFP-based cell culture reporter systems

Cell culture studies have the advantage of working with a large uniform population that can easily be manipulated. Transgene expression is possible after transient transfection of an expression vector or stably by selection of clonal transgenic cell lines. *Drosophila* Schneider 2 (S2) cells (Schneider, 1972) have the additional benefit of easy RNAi treatment by soaking, that is the uptake of dsRNA triggers that have simply been added to the medium.

GFP-based reporter cell lines allow the study of small RNA silencing pathways quantitatively by flow cytometry. Figure 3 shows the underlying principle using the example of a microRNA reporter. The reporter expresses GFP mRNA with specific miRNA binding sites in its 3´UTR (Figure 3A). Due to constitutive silencing by the endogenously expressed miRNA, reporter cells are only marginally green fluorescent owing to translational repression and degradation of the GFP message (Figure 3B). Any disturbances of the miRNA silencing pathway will increase GFP levels while enhanced silencing will be mirrored by lower GFP levels. Similar GFP-based systems are conceivable for other small RNA silencing pathways.

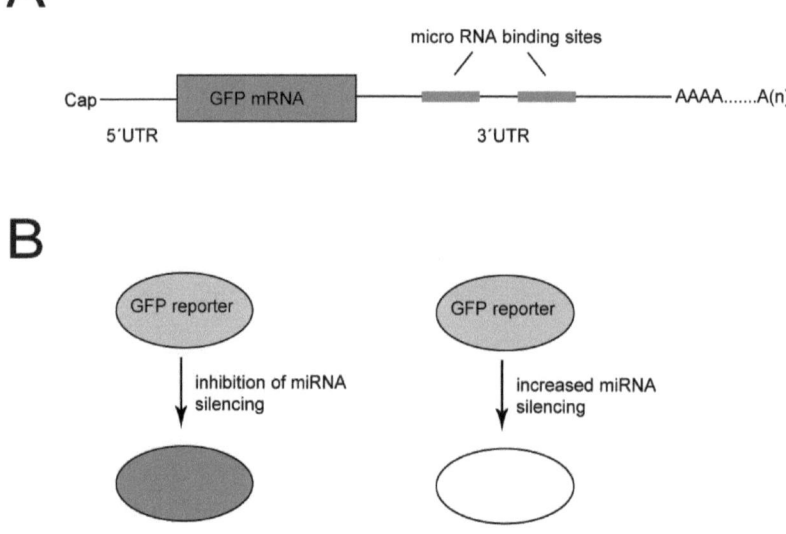

Figure 3: Schematic overview of GFP-based reporter systems
A) GFP-mRNA with multiple small RNA binding sites in its 3´UTR; as an example a miRNA-dependent reporter is depicted
B) Manipulation of pathway efficiency is directly visible in the GFP level of cells; fluorescence levels can be quantified by flow cytometry

2.4.2 Transgenic flies

2.4.2.1 Embryo injection and transgenic fly lines

P-elements have colonized the *Drosophila* genome within 80 years and spread over wildtype populations world-wide. However, laboratory strains have been protected from invasion and are free of transposase activity necessary to mobilize P-elements. This fact has made P-elements an invaluable tool in the generation of transgenic flies that consists of two components (Figure 4). A gene of interest can be flanked by P-element direct repeat sites required for recognition by the transposase and integration into the genome. Without the transposase activity, however, this element is unable to mobilize itself. A second plasmid encodes for the transposase enzyme activity but lacks P-element direct repeat recognition sites. This means that the plasmid-encoded transposase activity can mobilize and integrate the gene of interest into the genome but will itself be lost in subsequent cell divisions. By injection of both plasmids into the germline cells of an early *Drosophila* embryo, it is possible to obtain adults that will bear stably transgenic heterozygous offspring.

2.4.2.2 The UAS/Gal4 expression system

A frequently used conditional expression system in Drosophila is derived from yeast (Figure 5). If a transgene is brought under the control of a DNA sequence recognized by the yeast transcription factor Gal4, called Upstream Activating Sequence (UAS), it is inactive in the fly due to the lack of the yeast transcription factor. This is especially helpful if expression of the transgene is harmful to the organism, since it allows generating stably transgenic lines. Expression of the gene of interest can then be turned on in a tissue-specific manner by mating the UAS-strain to a driver line which expresses the yeast transcription factor Gal4 under a tissue specific promoter. Effects of transgene expression can then be studied in the affected tissue of the offspring.

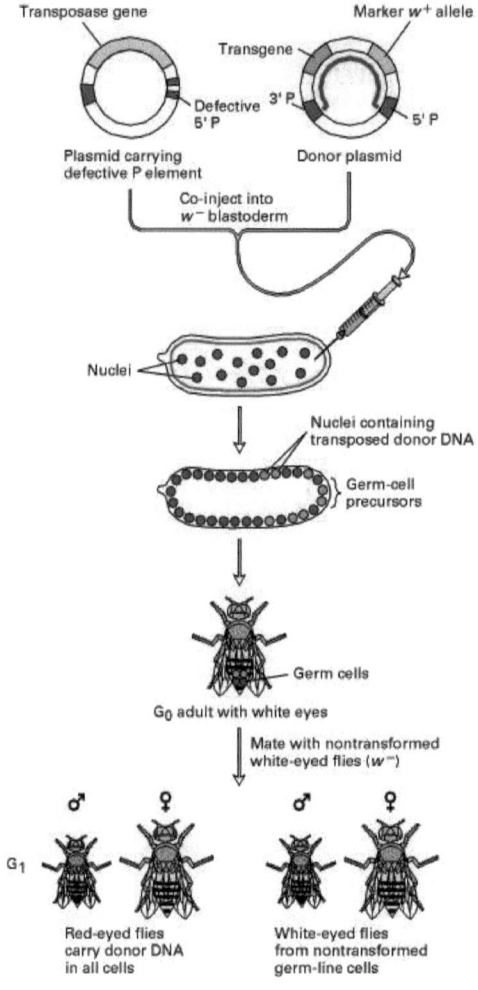

Figure 4: Generation of transgenic fruit flies by P-element transformation (legend continued on p. 22)

(legend Figure 4 continued)
The P element, a mobile genetic element, can move from one place in the genome to another. This movement (transposition) is catalyzed by transposase, which is encoded by the P element; the 3' and 5' ends of the P element are recognized by transposase and are required for transposition to occur. To produce transgenic fruit flies by this method, the functionally different regions of the P element are incorporated into two different bacterial plasmids. The donor plasmid contains three necessary elements: the transgene (orange); a marker gene (green) used to indicate flies in which the plasmid DNA is transposed to a recipient chromosome; and both ends of the P element (dark purple) — 3' P and 5' P — flanking the other two genes. It does not contain transposase. In this example, the marker is the dominant w^+ allele, which confers red eye color. The red bracket indicates the segment of the donor plasmid that can transpose into the fly genome. The other plasmid carries the P element (encoding transposase) with mutations in one end, which prevent it from transposing. The two plasmids are co-injected into blastoderm embryos homozygous for the recessive w^- allele, which confers white eye color. Transposase synthesized from the gene on the P-element plasmid catalyzes transposition of the donor plasmid DNA into the fly genome. Because transposition occurs only in germ-line cells (not in somatic cells), all the G_0 adults that develop from injected embryos have white eyes. Mating of these flies with white-eyed flies will yield some G_1 red-eyed progeny carrying the transgene and the marker allele (w^+) in all cells.
Figure from Lodish et al., 2000

2.5 Small RNA silencing systems in other organisms

Even though small RNA silencing pathways are conserved between most eukaryotes they differ in mechanistic details. Mammals only have a single Dicer molecule for both miRNAs and siRNAs (Grishok et al., 2001; Hutvagner et al., 2001; Ketting et al., 2001; Knight et al., 2001). However, mammalian genomes encode for two dsRBPs that interact with Dicer, TRBP and PACT (Chendrimada et al., 2005; Haase et al., 2005; Lee et al., 2006). These highly conserved homologs of *Drosophila* Loqs and R2D2, respectively, could similarly achieve commitment of the single Dicer to either the miRNA or the siRNA pathway. Like Loqs, its homolog TRBP is expressed in several splice variants (Haase et al., 2005). *C. elegans* has one canonical Dicer protein and three Dicer-related helicase proteins (DRHs) that are involved in small RNA mediated gene silencing (Tabara et al., 2002; Nakamura et al., 2007). The dsRBP RDE-4 is a *C. elegans* homolog of Loqs and R2D2 (Tabara et al., 2002).

Unlike mammals or flies, *C. elegans* is able to amplify and maintain an siRNA response for a long time and even transmit this silencing activity to the next generation (reviewed in Ghildiyal et al., 2009). The worm expresses 27 distinct Argonaute proteins compared with five in the fly. Most of these are specialized to bind secondary siRNAs that are derived from RNA-dependent RNA-polymerase (RdRP) activity (Aoki et al., 2007; Pak et al., 2007; Sijen et al., 2007).

Plants have developed a great diversity of small RNA silencing pathways and corresponding proteins which may be a way of an immobile organism to counteract various biotic and abiotic threats (reviewed in Ghildiyal et al., 2009). Small RNA mediated gene silencing in plants is characterized by the importance of transcriptional gene silencing on DNA and chromatin level. Indications for this epigenetic control by DNA and histone methylation were reported early in the field of small RNA-dependent silencing (Kooter et al., 1999; Mette et al., 2000).

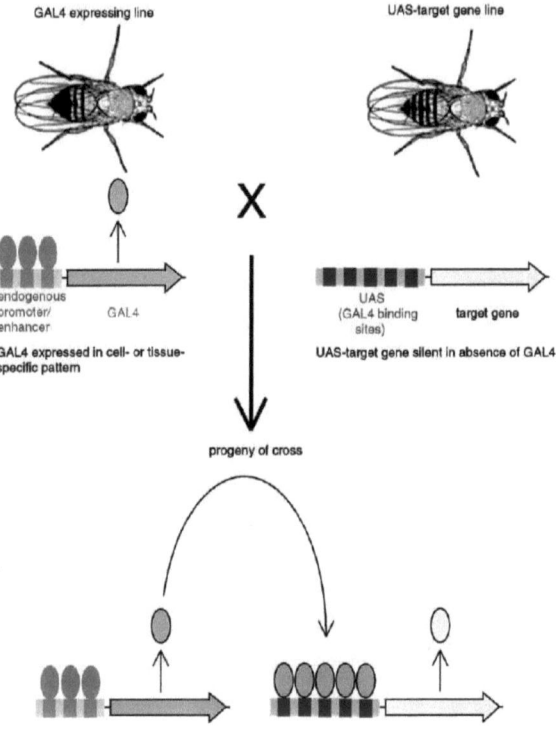

Figure 5: The UAS-Gal4 expression system in *Drosophila melanogaster*
Flies carrying a transgene with a UAS-promotor element are mated with a diver line, encoding for the yeast transcription factor Gal4. In the offspring, the gene of interest is activated.
Adapted from Alcorta et al., 1996

3 Specific aims of this thesis

1.) Which isoform of Loqs is involved in endo-siRNA dependent silencing? Can involvement of different isoforms in miRNA silencing and endo-siRNA silencing explain how specificity can be maintained despite involvement of Loqs in both processes?

2.) Can this particular isoform of Loqs interact with Dcr-2? If so, can the interaction domains be mapped more precisely?

3.) How can Dcr-2 bind both to R2D2 and Loqs? Does binding of one dsRBP influence binding of the other?

4.) What conclusion can be drawn for mechanisms of small RNA sorting in *Drosophila melanogaster*?

4 Materials and Methods

4.1 Materials

4.1.1 Laboratory hardware

ABI PRISM 7000 qPCR cycler	Applied Biosystems; Foster City, USA
Agarose gel running chamber (H1-Set)	Carl Roth GmbH; Karlsruhe, Germany
Branson Sonifier 250	Heinemann Ultraschall Labortechnik; Schwäbisch-Gmünd, Germany
FACSCalibur flow cytometer	Becton, Dickinson; Franklin Lakes, USA
Flow buddy CO_2-distributer	Genesee Scientific; San Diego, USA
Fly anesthetic pad/pistol	Genesee Scientific; San Diego, USA
INTAS UV Imaging System	INTAS; Göttingen, Germany
LAS 3000 mini Western Imager	Fujifilm; Tokyo, Japan
Leica MZ7 stereomicroscope	Leica Microsystems; Wetzlar, Germany
PAGE-electrophoresis material	BioRad; Hercules, USA
Power supply for electrophoresis	BioRad; Hercules, USA
Rotanta 460R centrifuge	Hettich; Tuttlingen, Germany
Semi-dry blotter	BioRad; Hercules, USA
SpectroLinker XL1500 UV Crosslinker	Spectronics Corporation; Westbury, USA
SterilGARD cell culture workbench	The Baker Company; Sanford, USA
Super Roller	Kisker; Steinfurt, Germany
Table top centrifuge (5417R and 5415R)	Eppendorf AG; Hamburg, Germany
Tank-blotting chamber	BioRad; Hercules, USA
Thermocycler	Eppendorf AG; Hamburg, Germany
Typhoon 9400 Variable Mode Imager	GE Healthcare; Freiburg, Germany

4.1.2 Analysis software

BD Cell Quest	Becton, Dickinson; Franklin Lakes, USA
BioEdit	see Hall, 1999
Multi Gauge V3.0	Fujifilm; Tokyo, Japan

4.1.3 Laboratory chemicals

Acrylamide	Carl Roth GmbH; Karlsruhe, Germany
Agarose	Biozym Scientific GmbH; Oldendorf, Germany
Ampicillin	Carl Roth GmbH; Karlsruhe, Germany
APS (ammonium peroxodisulfate)	Carl Roth GmbH; Karlsruhe, Germany
Bacto™ Agar	Becton, Dickinson; Franklin Lakes, USA
Bradford Assay	BioRad; Hercules, USA
BSA (bovine serum albumin)	Fermentas; St. Leon-Rot, Germany
Chloroform	Merck Biosciences GmbH; Schwalbach, Germany
Complete® without EDTA (=protease inhibitor cocktail)	Roche Diagnostics; Mannheim, Germany
Coomassie Brilliant Blue R	Sigma Aldrich; Taufkirchen, Germany
Coomassie G250	Carl Roth GmbH; Karlsruhe, Germany
Desoxyribonucleotides (dA/C/G/TTP)	Sigma Aldrich; Taufkirchen, Germany
DMSO (dimethyl sulfoxide)	Carl Roth GmbH; Karlsruhe, Germany
DTT (dithiothreitol)	Carl Roth GmbH; Karlsruhe, Germany
Ethanol (p.a.)	Merck Biosciences GmbH; Schwalbach, Germany
Ethanol (tech.)	VWR; Ismaning, Germany
FACS Flow/Clean/Rinse	Becton, Dickinson; Franklin Lakes, USA
Fetal bovine serum (FBS)	Thermo Fisher Scientific; Waltham, USA
Formamide	Sigma Aldrich; Taufkirchen, Germany
Fugene® HD transfection reagent	Roche Diagnostics GmbH; Mannheim, Germany
G418 sulphate (neomycin)	PAA, The Cell Culture Company; Cölbe, Germany
H_2O HPLC quality	VWR; Ismaning, Germany

Hepes (N-(2-Hydroxyethyl-)piperazine-N´-(2-ethanesulfonic acid))	Carl Roth GmbH; Karlsruhe, Germany
Hygromycin B	Carl Roth GmbH; Karlsruhe, Germany
IPTG	Fermentas; St. Leon-Rot, Germany
Isopropanol (p.a.)	Merck Biosciences GmbH; Schwalbach, Germany
Kanamycin	Carl Roth GmbH; Karlsruhe, Germany
L-Glutathione reduced	Sigma Aldrich; Taufkirchen, Germany
Methanol (p.a.)	Merck Biosciences GmbH; Schwalbach, Germany
Methanol (tech.)	VWR; Ismaning, Germany
NP-40 (Igepal CA 630)	Fluka BioCemika; Ulm, Germany
Powdered milk, Rapilait	Migros; Zürich, Switzerland
Roti®Aqua Phenol/C/I	Carl Roth GmbH; Karlsruhe, Germany
Saponin	Fluka BioCemika; Ulm, Germany
SDS (sodium dodecyl sulfate)	Merck Biosciences GmbH; Schwalbach, Germany
Sequagel Sequencing System	National Diagnostics; Atlanta, USA
Syber Safe/Gold	Invitrogen; Karlsruhe, Germany
TEMED (N,N,N´,N´-Tetramethylethylenediamine)	Carl Roth GmbH; Karlsruhe, Germany
Triton X-100	Sigma Aldrich; Taufkirchen, Germany
Trizol	Invitrogen; Karlsruhe, Germany
Tween 20	Carl Roth GmbH; Karlsruhe, Germany

Other standard laboratory chemicals were obtained from the in-house supply system.

4.1.4 Radiochemicals

[$\gamma\,^{32}$P] ATP (SRP 501) 10 mCi/ml; 6000 Ci/mmol; 250 µCi	Hartmann Analytic; Braunschweig, Germany

4.1.5 Enzymes
4.1.5.1 General enzymes

DNAse I, RNAse free	Fermentas; St. Leon-Rot, Germany
Polynucleotidekinase (PNK) with Buffer A	Fermentas; St. Leon-Rot, Germany
Proteinase K	Fermentas; St. Leon-Rot, Germany
RNAse A	Fermentas; St. Leon-Rot, Germany
T4-DNA Ligase	New England Biolabs; Ipswich, USA

4.1.5.2 Polymerases

Pfu DNA Polymerase	Fermentas; St. Leon-Rot, Germany
Phusion Hot Start DNA Polymerase	Finnzymes via New England Biolabs
Superscript II, Reverse Transcriptase	Invitrogen; Karlsruhe, Germany
T7-polymerase	laboratory stock
Taq DNA Polymerase	laboratory stock

4.1.5.3 Restriction enzymes

Fermentas; St. Leon-Rot, Germany and
New England Biolabs; Ipswich, USA

*Avr*II
*Bam*HI
*Bgl*II
*Eco*RI
*Kpn*I
*Nhe*I
*Not*I
*Hin*P1I
*Msp*I

4.1.6 Kits

DyNAmo Flash SYBR Green qPCR Kit	Finnzymes via New England Biolabs
GeneRACER Kit	Invitrogen; Karlsruhe, Germany
miScript SYBR Green PCR Kit	Qiagen; Hilden, Germany
QIAGEN Gel extraction Kit	Qiagen; Hilden, Germany

QIAGEN miRNeasy Mini-Kit	Qiagen; Hilden, Germany
QIAGEN PCR Purification Kit	Qiagen; Hilden, Germany
QIAGEN Plasmid Midi Kit	Qiagen; Hilden, Germany
QIAGEN Plasmid Mini Kit	Qiagen; Hilden, Germany
CloneJet PCR Cloning Kit (TA-cloning)	Fermentas; St. Leon-Roth, Germany

4.1.7 Other materials

cell culture materials	Bio&Sell; Nürnberg, Germany
	Sarstedt; Nümbrecht, Germany
Cryovials	Biozym Scientific GmbH; Oldendorf, Germany
ECL substrate	Thermo Fisher Scientific; Waltham, USA
Gene Ruler DNA Ladder Mix	Fermentas; St. Leon-Rot, Germany
GFP-Trap®_A beads	Chromotek; Planegg-Martinsried, Germany
Glutathione Sepharose 4 Fast Flow	Amersham Biosciences via GE Healthcare
Isopropanol freezing container	Nalgene via Thermo Fisher Scientific
NHS-Sepharose (H8280)	Sigma Aldrich; Taufkirchen, Germany
Nitrocellulose membrane (Protan BA 83)	Schleicher & Schüll; Dassel, Germany
Nylon membrane, positively charged	Roche Diagnostics GmbH; Mannheim, Germany
Parafilm	Carl Roth GmbH; Karlsruhe, Germany
Phosphoimager Screen	Fujifilm; Tokyo, Japan
Pistils for fly lysis	Sigma Aldrich; Taufkirchen, Germany
Polyvinylidenfluoride (PVDF) membrane	Milipore; Billerica, USA
Prestained Protein Ladder	Fermentas; St. Leon-Rot, Germany
Protein G Plus/Protein A Agarose beads (IP05)	Calbiochem via Merck
qPCR plates	Biozym Scientific GmbH; Oldendorf, Germany
RestoreTM Western Blot Stripping Buffer	Thermo Fisher Scientific; Waltham, USA

Sephadex spin column (G25)	Roche Diagnostics GmbH; Mannheim, Germany
Spin column for IP	MoBiTec; Göttingen, Germany
SuperSignal West Dura Extended Duration	Thermo Fisher Scientific; Waltham, USA
Whatman 595 ½ Folded Filters	Whatman GmbH; Dassel, Germany
Blotting paper	Machery-Nagel; Düren, Germany
α-Flag affinity agarose (A2220)	Sigma Aldrich; Taufkirchen, Germany
α-myc affinity agarose (A7470)	Sigma Aldrich; Taufkirchen, Germany

4.1.8 Plasmids in laboratory stock

Plasmid name	description	comments	reference
pGEX-6P-1	ampicillin resistance	for recombinant expression of GST-tagged proteins	Amersham/GE Healthcare
pHShygro	hygromycin resistance	selection of stable cell culture lines	
pHSneo	neomycin resistance	selection of stable cell culture lines	
pJB17	contains 2 kb tubulin promotor	excised with EcoRI/KpnI for cloning	Brennecke et al., 2003
pKF109	myc-Loqs-PB		Forstemann et al., 2005
pKF111	myc-Loqs-PA		Forstemann et al., 2005
pKF112	myc-Loqs-PB C-term.		Hartig et al., submitted
pKF114	myc-Loqs-PD$_{genomic}$		Hartig et al., submitted
pKF63	myc-GFP		Forstemann et al., 2005

pKF95	UAS-myc-GFP		Forstemann et al., 2005
pKF201	Flag-myc-tag only	also contains His$_6$-tag and Profinity-exact® tag	Hartig et al., submitted
pKF202	Flag-myc-Loqs-PB	also contains His$_6$-tag and Profinity-exact® tag	Hartig et al., submitted
pKF203	Flag-myc-Loqs-PA	also contains His$_6$-tag and Profinity-exact® tag	Hartig et al., submitted
pKF204	Flag-myc-Loqs-PB C-term.	also contains His$_6$-tag and Profinity-exact® tag	Hartig et al., submitted
pKF205	Flag-myc-Loqs-PD$_{genomic}$	also contains His$_6$-tag and Profinity-exact® tag	Hartig et al., submitted
pET-28a	kanamycin resistance	for recombinant expression of His$_6$-tagged proteins	Novagen/Merck Biochemicals
pUC18		transfection control	Stratagene; La Jolla, USA
pUAST		conditional expression	Brand et al., 1993
pCasper5	cloning vector	for Flag-Dcr-2 cloning	Le et al., 2007

4.1.9 Cells

4.1.9.1 Bacterial stocks

XL2-blue CaCl$_2$-competent cells Laboratory stock

BL21 Gold (DE3; pLys S) *E. coli* expression strain Stratagene; La Jolla, USA

SoloPack Gold Competent Cells Stratagene; La Jolla, USA

4.1.9.2 Cell lines

Cell line	Description	Comment	Reference
63N1	myc-GFP; no miRNA binding sites	endo-siRNA cell culture reporter	Hartig et al., 2009
63-6	myc-GFP; no miRNA binding sites	endo-siRNA cell culture reporter with less pronounced reaction	Förstemann, 2007; Hartig et al., 2009
67-1D	myc-GFP; two perfect binding sites for miR-277 in GFP 3'-UTR	miR-277 perfect match reporter for Dcr-1/Loqs-PB dependent biogenesis and Dcr-2/R2D2 dependent loading into Ago2	Förstemann, 2007
S2 B2	parental cell line	laboratory stock	

4.1.10 Fly stocks

Bloomington Stock #	genotype	description	origin
	w/w; P{w+, UASP-Loqs-PB}/CyO	UAS-Loqs-PB	Park et al., 2007
BL5138	y1 w*; P{w+mC=tubPGAL4} LL7/TM3, Sb1	tubulin-Gal4 driver line	Bloomington Stock Center
	yw/yw; 63-1/63-1; +/+	myc-GFP	Forstemann et al., 2005
	yw, hs-FLP/yw, hs-FLP/p{w+, loqs^{KO2-48}}, FRT40A/CyO; p{w+, Loqs-L (=PB)}$^{298-ba}$/TM3	Loqs-PB rescue	Park et al., 2007
BL7199	w*; Kr^{If-1}/CyO; D^1/TM6C, Sb^1 Tb^1	double-balancer	Bloomington Stock Center
	Loqs^{f0079}	Loqs promotor transposon insertion	Forstemann et al., 2005
BL6326	w^{1118}	recessive white mutation	Bloomington Stock Center

4.1.11 PCR primers

4.1.11.1 Cloning

4.1.11.1.1 Loqs / R2D2 constructs

Loqs_myc_PB_RB1_BamHI_fw	5'-AGCGGATCCATGGAACAAAAACTTATTTCTGAAGAAGACTTGGAACAAAAACTTATTTCTGAAGAAGACTTGGCCAAGAACACCATGGACCAGGAG-3'
Loqs_PB_RB1_BamHI_fw	5'-CAAGGATCCAAGAACACCATGGACCAGGAG-3'
Loqs_PB_RB1_BglII_rv	5'-CAAAGATCTGGCATTGCCGTCTCCTCCGCT-3'
Loqs_PB_RB2_BamHI_fw	5'-CAAGGATCCAATGCCACAGGCGGAGGAGAT-3'
Loqs_PB_RB2_BglII_rv	5'-CAAAGATCTTAACTTAAGCAGTTTTTTGCCCGTTGC-3'
Loqs_PB_RB3_BamHI_fw	5'-CAAGGATCCCAGAAGACTTGCTTGAAGAACAACAAG-3'
Loqs_PB_RB3_BglII_rv	5'-CAAAGATCTAGTTGGCTGCACCCCCTACTT-3'
Loqs_PB_RB3_NotI_rv	5'-CAAAGCGGCCGCTGGCTGCACCCCCTACTT-3'
Loqs_PA_RB2_BglII_rv	5'-CAAAGATCTTTCGCCCTCCAACTCGCCGCAG-3'
Loqs_PA_RB2_stop_Not_rv_new	5'-CAAAGCGGCCGCCTATTCGCCCTCCAACTCGCCGCAG-3'
Loqs_PB_RB2_NotI_rv	5'-CAAAGCGGCCGCTAACTTAAGCAGTTTTTTGCCCGTTGC-3'
Loqs_PB_RB2_stop_NotI_rv	5'-CAAAGCGGCCGCCTATAACTTAAGCAGTTTTTTGCCCGT-3'
BamHI_LoqsPC_fw	5'-CAAGGATCCAACGAATCTGTAAAGCACCT-3'
LoqsPC_end_NotI_rv	5'-CAAAGCGGCCGCCTACTGCGGGGCTGTAAATAAG-3'
BamHI_LoqsPD_fw	5'-CAAGGATCCGTGAGTATCATTCAAGACATC-3'
LoqsPD_end_NotI_rv	5'-CAAAGCGGCCGCTTAGATCTTGATGAACTC-3'
BamHI_myc_LoqsPDspec_fw	5'-AGCGGATCCATGGAACAAAAACTTATTTCTGAAGAAGACTTGGAACAAAAACTTATTTCTGAAGAAGACTTGGCCGTGAGTATCATTCAAGACATC-3'

R2d2_myc_RB1_BamHI_fw	5'-AGCGGATCCATGGAACAAAAACTTATTTCTGAAGAAGAC TTGGAACAAAAACTTATTTCTGAAGAAGACTTGGCCCTTGA ACTCATGGATAACAAG-3'
R2d2_RB1_BamHI_fw	5'-CAAGGATCCCTTGAACTCATGGATAACAAG-3'
R2D2_RB1_BglII_rv	5'-CAAAGATCTGTTGAGGTTAGTCAGTTCCTC-3'
R2d2_RB2_BamHI_fw	5'-CAAGGATCCCGGGACATGGTGAAGGAGCTG-3'
R2d2_RB2_BglII_rv	5'-CAAAGATCTTTTCAGGGTAGGATAGAAGTTCTTGAA-3'
R2_stop_Not_rv	5'-CAAAGCGGCCGCCTATTTCAGGGTAGGATAGAAGTT CTT-3'
R2d2_Rc_BamHI_fw	5'-CAAGGATCCAAGGAGGCCATTGAGGCCATC-3'
R2d2_Rc_BglII_rv	5'-CAAAGATCTATGTTATACGCATTAAATCAA-3'
R2d2_Rc_NotI_rv	5'-CAAAGCGGCCGCTTATACGCATTAAATCAA-3'

4.1.11.1.2 Flag-myc-Dcr-2 constructs

Nhe_tub3UTR_fw	5'-CAAGCTAGCATTCGAATCGGAAATCAATCGAATTC-3'
tub3UTR_AvrII_rv	5'-CAACCTAGGAGACTTGTGAACAAAATTGGATCCG-3'

NotI_Dcr2_AUG_fw	5'-CAAGCGGCCGCATGGAAGATGTGGAAATCAAGCCT CGC-3'
NotI_Dcr2_AUG_Flag_fw	5'-CAAGCGGCCGCATGGATTATAAAGATGATGATGATAAA GAAGATGTGGAAATCAAGCCTCGC-3'
Dcr2_stop_NheI_rv	5'-CAAGCTAGCTTAGGCGTCGCATTTGCTTAGCTGCTG-3'
D_hel_Flag_Dcr2_fw	5'-CAAGCGGCCGCATGGATTATAAAGATGATGATGATAAA CAGGACGATATTGACCCTTTTACCA-3'
D_RBD_Dcr2_stop_NheI_rv	5'-CAAGCTAGCTTAGAAAATTACCTCCCAAGTACGTTGG AG-3'

4.1.11.1.3 GFP-fusion constructs

BamHI_myc_GFP_fw	5'-AGCGGATCCATGGAACAAAAACTTATTTCTGAAGAAGACTTGGAACAAAAACTTATTTCTGAAGAAGACTTGGCCGTGAGCAAGGGCGAGGAGCTG-3'
BglII_GFP_ORF_rv	5'-CAAAGATCTCTTGTACAGCTCGTCCATG-3'

4.1.11.1.4 T7-primers for RNAi

T7prom_Loqs 5´UTR_fw	5'-CGTAATACGACTCACTATAGGGCAACCACAAATATCAGT-3'
T7prom_Loqs 5´UTR_rv	5'-CGTAATACGACTCACTATAGGTTGCACGGTTTTCGGGAG-3'
T7prom_Loqs 3´UTR_fw	5'-CGTAATACGACTCACTATAGGGTGCAGCCAACTGAATAGCA-3'
T7prom_Loqs 3´UTR_rv	5'-CGTAATACGACTCACTATAGGCCTTCGCAAACTAGCACGTAG-3'
T7prom_R2D2 3´UTR_fw	5'-CGTAATACGACTCACTATAGGATTCAACTATTCTAGCTTA-3'
T7prom_R2D2 3´UTR_rv	5'-CGTAATACGACTCACTATAGGCTTTGATTACTAGCATTCCT-3'
T7prom_GFP_ORF_fw	5'-CGTAATACGACTCACTATAGGATGGTGAGCAAGGGCGAGGAGCTG-3'
T7prom_GFP_ORF_rv	5'-CGTAATACGACTCACTATAGGTTACTTGTACAGCTCGTCCATG-3'
T7prom_LoqsPC_fw	5'-CGTAATACGACTCACTATAGGAACGAATCTGTAAAGCACCT-3'
T7prom_LoqsPC_rv	5'-CGTAATACGACTCACTATAGGCTGTAAATAAGAGCGCAAAG-3'
T7prom_Dcr1_new_fw	5'-CGTAATACGACTCACTATAGGCCAGGATCAACCGCAGTATT-3'
T7prom_Dcr1_new_rv	5'-CGTAATACGACTCACTATAGGGTATCGTGGCGTGAGGAAGT-3'

T7prom_LoqsPB+PC_fw	5'-CGTAATACGACTCACTATAGGCCCCGCAGTAGTGAAAATTA-3'
T7prom_LoqsPB+PC_rv	5'-CGTAATACGACTCACTATAGGCTGTAACTTAAGCAGTTTTTTGCC-3'
T7prom_HP1_ORF_fw	5'-CGTAATACGACTCACTATAGGATGGGCAAGAAAATCGACAA-3'
T7prom_HP1_ORF_rv	5'-CGTAATACGACTCACTATAGGACCATTTCTGCTTGGTCCAC-3'
T7prom_LoqsPD_fw	5'-CGTAATACGACTCACTATGTGAGTATCATTCAAGACATCGATC-3'
T7prom_LoqsPD_rv	5'-CGTAATACGACTCACTATAGGTAAGGTGTAAGCATTATGTTAATT-3'
T7prom_dsRed_ORF_rv	5'-CGTAATACGACTCACTATAGGTGGTGTAGTCCTCGTTGTGG-3'
T7prom_dsRed_ORF_rv_short	5'-CGTAATACGACTCACTATAGGCCGTCCTCGAAGTTCATCAC-3'

4.1.11.2 Sequencing

4.1.11.2.1 Loqs sequencing primer

Seq_L1_rv	5'-GATATCTTCTCCTTCTTGAGCTTCACATGG-3'
Seq_L1-L2_fw	5'-CAGGACGATCGAAGAAGGAGGCCAAG-3'
Seq_L1-L2_rv	5'-TGCCCTTGCCCATCTCGCGGTA-3'
Seq_L2_L3_fw	5'-CATCGATTCGGGCAAAATCAGCGACAG-3'
Seq_Loqs_L2-L3_rv	5'-GAACTGGTTCTCCGTGGCGATTTC-3'
Seq_Loqs_L3_fw	5'-GGCAGCGGACCAACAG-3'

4.1.11.2.2 R2D2 sequencing primer

Seq_R2D2_R1_rv	5'-CGGGAATAAAACTGTATGTTGGTAG-3'
Seq_R2D2_R1-R2_fw	5'-CGGCATACACGGCTTGATGAAG-3'
Seq_R2D2_R1-R2_rv	5'-CCACGGAGCAACAGGCCACGAAT-3'
Seq_R2D2_R2-R3_fw	5'-CGTAGACCATACAGGCATGCGG-3'
Seq_R2D2_R2-R3_rv	5'-GTGTTATCTTTAAAGAGCTCATTACGTC-3'
Seq_R2D2_R3_fw	5'-GCGTCGAGCTTAATTGTGCATTC-3'

4.1.11.2.3 Flag-Dcr-2 sequencing primer

pCasper5_seq_fw	5'-TGGTACATCAAATACCCTTGGATCG-3'
pCasper5_seq_rv	5'-GTGCGAGTGAAAGGAATAGTATTCTG-3'

4.1.11.3 qPCR

qPCR loqs fw 2	5'-AATGCCGTCAGTGGTAGTCC-3'
qPCR Loqs rv 2	5'-CGTTTCGCTGACGAACTTTA-3'
qPCR Probe B (*CG4068B*)	5'-TTGACTCCAACAAGTTCGCTCC-3'

4.1.11.4 Test-PCR

Test_iso_PD_rv	5'-AGCATGGGACTGCATTCAA-3'
Loqs_all_fw	5'-CAAAATCAGCGACAGCATCTGCGG-3'
Loqs_all_rv_long	5'-AACTGGTTCTCCGTGGCGATTTCG-3'
GFP 2 s	5'-AGAACGGCATCAAGGTGAAC-3'
GFP 2 as	5'-TGCTCAGGTAGTGGTTGTCG-3'

4.1.11.5 RACE

3´RACE exon3 fw = Loqs_PB_RB2_BamHI_fw	5´-CAAGGATCCAATGCCACAGGCGGAGGAGAT-3´
Cloned AMV RT Module	5'-GCTGTCAACGATACGCTACGTAACGGCATGACAGTG(T)$_{18}$-3'
3´primer	5'-GCTGTCAACGATACGCTACGTAACG-3'
3´nested primer	5'-CGCTACGTAACGGCATGACAGTG-3'

4.1.11.6 Mapping P-element insertions in transgenic flies

4.1.11.6.1 Sequencing primer for wht1/lac1

Sp1_seq_mapping 5´PCR	5'-ACACAACCTTTCCTCTCAACAA-3'
Spep1_seq_mapping 3´PCR	5'-GACACTCAGAATACTATTC-3'

4.1.11.6.2 Verification of mapped insertion site

UAS-Loqs-PD line „B": 3R insertion site 217147

3R:2171431..2171458; Forward:	5'-CACATGCCGCTGCCAGTTACGCCATTTC-3'
3R:2171468..2171485; Reverse:	5'-TCCCGGGGCAGATGGGAC-3'

UAS-Loqs-PC line „E": 3L insertion site 5241560

3L:5241514..5241540; Forward:	5'-GTCTCGTCGGGTCGAGCAACGAAGTTC-3'
3L:5241559..5241587; Reverse:	5'-CTCTAGACTCTGAATCTGAAACTGAAGTC-3'

UAS-Loqs-PD line „D": 2L insertion site 5999610

2L:5999562..5999586; Forward:	5'-CAAACACAGTTTCTTATCGGCGGAG-3'
2L:5999644..5999672; Reverse:	5'-CCTATGACCGAACTGATTTTGAATCTAATG-3'

UAS-Loqs-PD line „H": 3L insertion site 20488501

3L:20488438..20488468; Forward:	5'-CCTGTCAATTCTAGTATATTCCAATTGCTAC-3'
3L:20488535..20488561; Reverse:	5'-ACCAACTCGATTCGGTTAATTTACGCT-3'

UAS-Loqs-PC line „M": 2R insertion site 8519550

2R:8519501..8519524; Forward:	5'-CGGTAGATCTCAAAGTTCAGGCGC-3'
2R:8519574..8519599; Reverse:	5'-CTCGGCCCTTAATCCATCTTTAAACC-3'

4.1.11.7 Sequencing primer for pUASP-trangenic flies

pUASP basal Promotor	5'-ATTCAGTGCACGTTTGCTTG-3'

4.1.12 Media

4.1.12.1 Bacterial stocks

All *E. coli* strains were cultivated in LB-medium or in SOC-medium following transformation. Agarose plates were obtained from in-house supply.

SOB-medium	0.5% (w/v) yeast extract
	2% (w/v) Tryptone
	10 mM NaCl
	2.5 mM KCl
	10 mM $MgCl_2$
	10 mM $MgSO_4$
	pH 7
SOC-medium	SOB-medium
	20 mM glucose
LB-medium	1% (w/v) Tryptone
	0.5% (w/v) yeast extract
	1% (w/v) NaCl
	pH 7.2

Antibiotics added to medium after autoclaving:

100 µg/ml ampicillin (100 mg/ml stock)

10 µg/ml kanamycin (10 mg/ml stock)

25 µg/ml chloramphenicol (34 mg/ml stock)

4.1.12.2 Cell culture

Cell culture medium and additives for *Drosophila* Schneider cells was obtained from Bio&Sell (Nürnberg, Germany) and supplemented with 10% temperature-inactivated Fetal Bovine Serum (FBS; Thermo Fisher; Waltham, USA).

For selection purposes 1.2 mg/ml G418 (neomycin) or 300 µg/ml hygromycin were added to the medium.

4.1.13 Fly food
Standard fly food was obtained from in-house supply:

5.8% corn meal

5.5% molasses

2.4% yeast extract

4.1.14 Antibodies

4.1.14.1 Primary antibodies

antibody	organism	dilution	reference/ catalog #
α-beta-tubulin	mouse (monoclonal)	1:1000	DSHB, E7
α-Dcr-1	rabbit (polyclonal)	1:1000	Förstemann, 2007
α-Dcr-2	rabbit (polyclonal)	1:1000	Abcam, ab4732-100
α-Dcr-2	mouse (monoclonal)	1:1000	Dcr-2 8-59; Miyoshi et al., 2009
α-Flag	mouse (monoclonal)	1:2000	Sigma, F1804
α-GFP	mouse (monoclonal)	1:4000	Santa Cruz Biotechnology, B-2; sc-9996
α-GST	mouse (monoclonal)	1:1000	GE Healthcare, 27-4577-01
α-Loqs	mouse (monoclonal)	1:1000-1:4000	Hartig et al., 2009; Miyoshi et al., 2009

α-Loqs-PB C-term.	rabbit (polyclonal)	1:1000	Abcam, ab 24237
α-myc	mouse (monoclonal)	1:1000	Sigma, M4439
α-R2D2	rabbit (polyclonal)	1:5000	Abcam, ab14750

Rabbit IgG was purchased from Sigma (I5006).

4.1.14.2 Secondary antibodies

Antibody	Dilution	Origin
Goat Anti-Mouse IgG (H+L) HRP-coupled	1:100 000	Pierce (Thermo Scientific) 31160
Goat Anti-Rabbit IgG (H+L) HRP-coupled	1:100 000	Pierce (Thermo Scientific) 31210

4.1.15 Stock solutions and commonly used buffers

Acrylamide solution (Rotiphorese Gel 30)   Acrylamide : Bisacrylamide = 37.5 : 1

Alternative IP-buffer   Lysis buffer with 0.4% Triton X-100

Buffer A for fly DNA extraction
100 mM Tris/HCl, pH 7.5
100 mM EDTA
100 mM NaCl
0.5% SDS

Church buffer
1% (w/v) bovine serum albumine
1 mM EDTA

	0.5 M phosphate buffer
	7% (w/v) SDS
	pH 7.2
Colloidal Coomassie staining solution	50 g/l aluminum sulfate
	2% (v/v) H_3PO_4 (conc.)
	10% (v/v) 100% ethanol
	0.5% (v/v) Coomassie G250 stock solution
Coomassie G250 stock solution	0.5 g/l Coomassie G250 in 100% methanol
Coomassie staining solution	45% (v/v) methanol
	10% acetic acid
	0.25% (w/v) Coomassie Brilliant Blue
Coomassie destain	45% (v/v) methanol
	10% acetic acid
DNA loading buffer (6x)	0.25% (w/v) bromophenol blue
	0.25% (w/v) xylene cyanol
	30% (w/v) glycerol
Formamide loading dye (2x)	80% (w/v) formamide
	10 mM EDTA, pH 8
	1 mg/ml xylene cyanol
	1 mg/ml bromophenol blue
GST-purification binding/washing buffer	1xPBS
	0.5% (w/v) Saponin
	1x Complete* without EDTA (=protease inhibitor cocktail)

GST-purification elution buffer	50 mM Tris/HCl
	10 mM Glutathione
	0.1% Saponin
Laemmli SDS loading buffer (2x)	100 mM Tris/HCl, pH 6.8
	4% (w/v) SDS
	20% (v/v) glycerol
	0.2% (w/v) bromophenol blue
	200 mM freshly added DTT
LiCl/KOAc Solution	1 part 5 M KOAc stock : 2.5 parts 6 M LiCl stock
Lysis buffer for protein extraction	100 mM KOAc
	30 mM Hepes
	2 mM $MgCl_2$
	1 mM DTT
	1% (v/v) Triton X-100
	2x Complete* without EDTA (=protease inhibitor cocktail)
Lysis buffer with high $MgCl_2$	20 mM Tris/Cl, pH 8
	20 mM $MgCl_2$
	0.5% NP-40
	2x Complete* without EDTA (=protease inhibitor cocktail)
PBS (10x)	137 mM NaCl
	2.7 mM KCl
	10 mM Na_2HPO_4
	2 mM KH_2PO_4, pH 7.4

PBS-T	PBS supplemented with 0.05% Tween-20
SDS-running buffer (5x)	125 mM Tris/HCl, pH 7.5 1.25 M glycine 5% SDS
SSC (20x)	3 M NaCl 0.3 M sodium citrate
TAE (50x)	2 M Tris-base 5.71% acetic acid 100 mM EDTA
TBE (10x)	0.9 M Tris base 0.9 M boric acid 0.5 M EDTA (pH 8)
TBS (10x)	50 mM Tris 150 mM NaCl pH 7.4
TBS-T	TBS supplemented with 0.02% Tween-20
Western blotting stock (10x)	250 mM Tris/HCl, pH 7.5 1.92 M glycine
Western blotting buffer (1x)	10% Western blotting stock (10x) 20% methanol

4.2 Methods

4.2.1 Molecular cloning

4.2.1.1 Primer design for cloning of dsRBDs

Loqs and R2D2 sequences were analyzed for evolutionary conserved dsRBD motifs with BLAST software (www.blast.ncbi.nlm.nih.gov; see Appendix 6 and Appendix 7). Primers were designed which annealed to the linker sequences and introduced appropriate restriction sites. Oligonucleotides were ordered from Thermo Fisher Scientific (Waltham, USA) and Eurofins/MWG (Ebersberg, Germany).

4.2.1.2 Amplification of DNA sequences by Polymerase Chain Reaction (PCR)

For molecular cloning purposes the standard reaction mix was as follows:

10x Taq-buffer (-MgCl$_2$, +(NH$_4$)$_2$SO$_4$)	5 µl
dNTP-mix (10 mM each)	1 µl
25 mM MgSO$_4$	3 µl (final conc. 1.5 mM)
100 nM primer fw	0.1 µl
100 nM primer rv	0.1 µl
template DNA	2 µl
Taq polymerase	0.4 µl
Pfu polymerase	0.1 µl
H$_2$O	<u>38.3 µl</u>
	50 µl

Reaction components were acquired from Fermentas (St. Leon-Rot, Germany), Taq polymerase was taken from our own laboratory stock. PCR reactions were carried out on an automated thermal cycler (Eppendorf; Hamburg, Germany).

The following standard protocol for gradient PCR was used to determine the appropriate annealing temperatures. Conditions were then adjusted accordingly:

3 min 94°C (initial denaturation)
34 cycles:
1 min 94°C (denaturation)

1 min 50-65°C (primer annealing)

1 min 72°C (extension)

5 min 72°C (final extension)

Hold 4°C

PCR products were separated by agarose gel electrophoresis, excised and purified by QIAGEN Gel Extraction Kit or directly treated with the QIAGEN PCR Purification Kit (Qiagen; Hilden, Germany).

4.2.1.3 Agarose gel electrophoresis

Appropriate for the length of nucleotides to be separated 0.5% – 4% agarose gels were prepared with 1x TAE buffer and stained with 1x SyberSafe (Invitrogen; Karlsruhe, Germany). Gels were run at 55V for 30 min and photographed in an Intas UV Imaging System. If higher sensitivity was required gels were re-stained in 1x SyberGold (Invitrogen; Karlsruhe, Germany) for 30 min.

4.2.1.4 Specific digestion of DNA by restriction endonucleases

Endonucleolytic digestion of DNA was carried out with restriction endonucleases acquired from Fermentas (St. Leon-Rot, Germany) and New England Biolabs (Ipswich, USA) according to manufacturers' recommendations. Usually, reactions were incubated for 2 hours at 37°C.

4.2.1.5 Ligation of vector with insert DNA

Digested and purified insert and vector were combined according to the following formula in a molar ratio of 1 to 3:

$$\frac{\text{mass (vector)} * \text{length bp (vector)} * 3}{\text{length bp (insert)}} = \text{required mass (insert)}$$

≥ 200 ng vector

+ required amount of insert

+ 2 µl T4-buffer (10x)

+ 1 µl T4-DNA Ligase

+ x µl H$_2$O

20 µl

Optimally samples were ligated over night at 18°C and used for bacterial transformation.

4.2.1.6 Bacterial transformation

Transformation of competent bacteria was carried out by standard heat shock procedures. Briefly, 50 µl XL2-blue $CaCl_2$-competent cells were thawed on ice. 1-4 µl of ligation sample were added and the mixture was incubated on ice for 30 min, subjected to a 1 min heat shock at 42°C and returned to ice. 1 ml SOC-medium was added and cells were allowed to grow for 1 h in a 37°C shaking incubator. Cells were then streaked out on agarose plates with appropriate antibiotics for selection of transformants.

4.2.1.7 Test for correct transformants by colony-PCR

Individual colonies were tested for correct integration of the insert by colony-PCR with suitable primer pairs. A standard PCR reaction mix was inoculated with a single colony, which was subsequently streaked onto a fresh plate and labeled for later recognition. Standard amplification was carried out with 10 min initial denaturing for cell lysis of bacteria.

4.2.1.8 Preparation of plasmid DNA

Plasmid DNA was prepared from over-night cultures of 5 ml or 30 ml LB-medium, respectively, supplemented with appropriate antibiotics (usually 100 µg/ml ampicillin). QIAGEN Mini or Midi Kits (Qiagen; Hilden, Germany) were used according to the manufacturer's protocols.

4.2.1.9 DNA sequencing

Sequencing was carried out by Eurofins/MWG (Ebersberg, Germany) according to the provider's specifications. Sequence analysis and alignments were performed with BioEdit software (Hall, 1999) and openly available online tools.

4.2.1.10 3´-RACE PCR analysis of the loqs-RD variant

RNA from S2-cells was prepared with the miRNeasy® mini-kit (Qiagen; Hilden, Germany). 1 µg of total RNA was reverse transcribed (Superscript II, Invitrogen; Karlsruhe, Germany) using the oligo(dT) primer from the GeneRACER Kit (Invitrogen; Karlsruhe, Germany) 5'-GCT GTC AAC GAT ACG CTA CGT AAC GGC ATG ACA GTG $(T)_{18}$-3'. PCR was performed with a primer at the end of *loqs* exon 3 (5'-CAA GGA TCC AAT GCC ACA GGC GGA GGA GAT-3') and the 3'-end reverse primer from the GeneRACER kit (5'-GCT GTC AAC GAT ACG CTA CGT AAC G-3') using a 5:1 mix of Taq and Pfu (Fermentas; St. Leon-Rot, Germany) polymerases and a 1:20 diluted cDNA template at 55°C annealing temperature. PCR products were TA-cloned using the CloneJet PCR cloning kit (Fermentas; St. Leon-Rot, Germany) and sequenced (Eurofins/MWG; Ebersberg, Germany).

4.2.2 Methods of *Drosophila* S2 cell culture

4.2.2.1 Maintenance

Drosophila Schneider 2 (S2) cells were cultured in Schneider's medium (Bio&Sell, Nürnberg, Germany) supplemented with 10% fetal calf serum (Thermo Fisher Scientific; Waltham, USA) in 10 cm cell culture dishes. Cultures were split 2-3 times a week into fresh medium.

4.2.2.2 Depletion of individual genes by RNAi in cell culture

4.2.2.2.1 Production of dsRNA

RNA interference by simply adding dsRNA to cell culture medium (soaking) was performed essentially as described (Shah et al., 2008). Briefly, gene specific primers for target genes were designed to introduce flanking T7-promotor fragments (see Materials 4.1.11.1.4) and amplified using standard PCR. The PCR products were precipitated with ethanol, the pellet was dissolved in 1/10 of the original volume and used directly for over-night in-vitro transcription at 37°C with the following specifications:

10 µl 10x T7-buffer
10 µl re-dissolved DNA
0.5 µl 1 M DTT
5 µl 100 mM ATP
5 µl 100 mM CTP
5 µl 100 mM UTP
8 µl 100 mM GTP
54.5 µl H_2O
<u>2 µl T7-polymerase</u>
100 µl

After in-vitro transcription 1 µl of DNAseI was added per 100 µl of reaction and incubated for 30 min at 37 °C. White precipitate was pelleted and RNA was precipitated from supernatant with 1x volume of isopropanol and washed with 70% ethanol. The dried pellet was dissolved in 100 µl of H_2O. For proper strand annealing $MgCl_2$ was added to a final concentration of 5 mM, the sample was heated to 95°C for 3 min and allowed to slowly cool down to room temperature. Concentration of dsRNA was estimated from an agarose gel in comparison to a DNA Ladder Mix (Fermentas; St. Leon-Rot, Germany).

4.2.2.2.2 Soaking

Cells were seeded at 0.5×10^6 cells/ml and 20 µg/ml dsRNA were added to the medium. On day 2, the cells were split 1:5 into a fresh culture dish and the dsRNA treatment was repeated. On day 5 or 6, GFP fluorescence was quantified in a Becton Dickinson FACSCalibur flow cytometer.

4.2.2.2.3 Transfection

Transfections of S2 cells were carried out essentially as described in Shah et al., 2008. For each well of a 24-well cell culture dish 50 ng of the vector of interest in 50 µl medium (without serum) and 4 µl of Fugene Transfection Reagent (Roche Diagnostics; Mannheim, Germany) in 46 µl of medium (without serum) were mixed and incubated for 1 hour. Cells were added to the transfection mix at 0.5×10^6 cells/ml medium (+10% FBS), split on day 3 after transfection and analyzed on day 5 or 6.

4.2.2.2.4 Flow cytometric analysis of GFP levels

GFP fluorescence levels were quantified using a FACSCalibur flow cytometer (Becton Dickinson; Franklin Lakes, USA). 100 µl of cell culture were added to 300 µl of FACS-flow. 10,000 cells were measured per sample (settings FL-1 290V Log scale). FACS profiles were depicted as cell counts over a logarithmic scale for fluorescence intensity. Analysis was carried out with CellQuest software (Becton Dickinson; Franklin Lakes, USA). Non-fluorescent GFP-reporter cells were excluded from the analysis and the mean fluorescence value for each sample was determined (Note: Stable cell-lines in non-selective medium may not be derived from a single clone and contain a minority of contaminating non-fluorescent cells). Measurements were carried out in technical triplicates to calculate mean and standard deviation.

4.2.2.3 Selection of clonal cell lines

To create cell lines that stably express a transgene the expression plasmid of interest was co-transfected with an antibiotic resistance plasmid into cells at $5\text{-}10 \times 10^5$ cells/ml. For native S2 cells 10 ng pHSneo (for neomycin resistance) and for secondary transfections of stable reporter lines 10 ng pHShygro (for hygromycin resistance) were used together with 200 ng of the vector of interest. After 3 days, cells were split 1:5 into G418 or hygromycin containing medium, respectively. The concentration was 1.2 mg/ml of G418 for neomycin resistance and 300 µg/ml of hygromycin. Cells were split 1:5 once a week for 4 weeks to obtain stable cell lines. For clonal selection serial dilution steps in a 96-well plate were made and colonies derived from a single cell were picked.

4.2.2.4 Storage of cells in liquid nitrogen

Cell stocks were frozen by adding 500 µl cells to 100 µl Dimethylsulfoxide (DMSO) diluted in 400 µl cell culture medium (+10% FBS) in a Cryovial (Biozym; Oldendorf, Germany). Cryovials were slowly (1°C per hour) cooled to -80°C in an isopropanol freezing container (Nalgene/Thermo Fisher) and transferred into liquid nitrogen for long-term storage.

4.2.3 Protein analysis

4.2.3.1 Protein extraction

Cells were harvested (2500 x g, 5 min) and washed twice in PBS. The pellet was resuspended in lysis buffer (30 mM Hepes, 100 mM KOAc, 2 mM $MgCl_2$, 1 mM fresh DTT, 1% (v/v) Triton X-100, 2x protease inhibitor cocktail (Complete* without EDTA, Roche Diagnostics)) and frozen in liquid nitrogen. Samples were thawed on ice and cell debris was pelleted in a refrigerated microcentrifuge at 15,000 x g (Eppendorf; Hamburg, Germany). Protein concentrations were determined by Bradford Assay (BioRad; Hercules, USA).

Fly protein was extracted by grinding flies in lysis buffer using a pistil (Sigma Aldrich; Taufkirchen, Germany) suitable for 1.5 ml reaction tubes and subsequent freeze-thaw lysis analogous to S2 cells.

For extraction of HP-1 protein (Figure 20B), lysis buffer with high $MgCl_2$ conditions was used as indicated [Tris/Cl 20 mM pH 8; $MgCl_2$ 20 mM; 0.5% NP40 + 2x protease inhibitor cocktail (Complete* without EDTA, Roche Diagnostics)].

4.2.3.2 Co-immunoprecipitation

Protein G Plus/Protein A Agarose beads (IP05, Calbiochem) were washed three times in 1 ml Lysis buffer and agitated for 30 min at 4°C with the respective antibody. Beads were then washed four times and incubated with 1-5 mg of total protein in Lysis buffer at 4°C for 1 h on an overhead-rotator. For α-myc immunoprecipitation in Figure 13A, B, C 1 mg total protein was incubated with 50 µl α-myc affinity agarose (A7470, Sigma; washed three times in 1 ml of Lysis buffer) for 2 h at 4°C on an overhead-rotator. For α-Flag immunoprecipitation, α-Flag affinity agarose (A2220, Sigma), washed three times in 1 ml of Lysis buffer, was used. For IP experiments in Figure 15B, C and Figure 19D a Triton X-100 concentration of 0.4% (v/v) was emloyed. GFP-fusion constructs were precipitated using GFP-Trap®_A beads (Chromotek; Planegg-Martinsried, Germany). Flow-through and beads were separated by spin columns (MoBiTec; Göttingen, Germany) and washed four times with 500 µl lysis buffer. Bound proteins were eluted by applying 15 µl 1x Laemmli SDS sample buffer and heating to 95°C for 5 minutes.

4.2.3.3 Immunoblotting for detection of proteins

Western blotting was performed as previously described (Förstemann, 2007). In short, proteins were separated on 8-12% polyacrylamide gels (150 V; 1h) in a BioRad electrophoresis tank. Proteins were transferred to a polyvinylidenfluoride (PVDF; Milipore; Billerica, USA) membrane by tank blotting (100 V; 1-1.5h). After blocking in 5% milk for 1 h membranes were incubated under constant rolling in 50 ml tubes with 5 ml of primary antibody solution over night at 4°C. Antibodies and dilutions are indicated in the Materials section 4.1.14.1. For all washing, blocking and incubation steps for rabbit antibodies PBS-T (0.05% Tween) was used, for mouse antibodies TBS-T (0.02% Tween) was employed. After primary antibody binding blotting membranes were washed three times 10 min in buffer and incubated with appropriate secondary antibodies for 2 h at room temperature. After analogous washing, Enhanced Chemiluminescence (ECL) substrate (Thermo Fisher Scientific; Waltham, USA) was applied and the signal was measured in an LAS3000 mini Western Imager System (Fujifilm; Tokyo, Japan). Multi Gauge software (Fujifilm; Tokyo, Japan) was used for relative quantification of protein band intensities. Western blots were stripped with 10 ml of Restore Stripping Solution (Thermo Fisher Scientific; Waltham, USA) for 30 min at 37°C and another 30 min at room temperature, washed extensively in water and buffer and blocked for new primary antibody incubation.

4.2.3.4 α-Loqs-PD-specific antibody production

Loqs-PD-specific peptide was synthesized, coupled to KLH (keyhole limpet hemocyanin) as a carrier protein and used to immunize rabbits (Davids Biotechnologie; Rgensburg, Germany). Affinity purification was performed as described (Harlow et al., 1988) using Loqs-PD peptide coupled to NHS-Sepharose (Sigma, H8280).

4.2.3.5 Dot blot

8-10 µg of total protein extracts were spotted on a nitrocellulose membrane (Schleicher & Schuell; Dassel, Germany) and left to dry. Immunoblotting was performed as described above.

4.2.4 RNA analysis

4.2.4.1 RNA extraction

RNA was extracted either with Trizol (Invitrogen; Karlsruhe, Germany) according to the manufacturer's instructions or using the QIAGEN miRNeasy Kit (Quiagen; Hilden, Germany) and quantified using spectrophotometry.

4.2.4.2 Northern Blotting

1-5 µg of RNA were separated on a 20% Sequagel Acrylamide/Urea gel (National Diagnostics; Atlanta, USA) at 250V for 1.5 hours. RNA was then transferred to a positively charged Nylon membrane (Roche Diagnostics; Mannheim, Germany) by semi-dry blotting for 1h at 20V. Membranes were transferred to hybridizing tubes and incubated in Church Buffer for at least 1h in a hybridization oven under constant rotation. Probes for detection of *bantam* miRNA and 2S rRNA were as described (Forstemann et al., 2005). A DNA antisense probe for the detection of *CG4068* B endo-siRNA (Okamura et al., 2008b) was used.

Probe for Northern Blotting	Sequence
Probe B (CG4068 B) DNA as probe	5´-GGAGCGAACTTGTTGGAGTCAA-3´
bantam 2´OMe as probe	5´-AATCAGCTTTCAAAATGATCTCA-3´
miR-277 2´OMe as probe	5´-TGTCGTACCAGATAGTGCATTTA-3´

Probes were labeled by incubating 9 µl H_2O, 2 µl 10x PNK buffer, 2 µl 5 mM probe oligonucleotide (=10 pmol), 1 µl PNK (Fermentas) and 6 µl [γ ^{32}P] ATP for 1h at 37°C. Unbound radioactive nucleotides were removed using a Sephadex G-25 spin column (Roche Diagnostics; Mannheim, Germany). Labeled oligonucleotide anti-sense probes were added to 5 ml of Church Buffer for over-night hybridization. Prehybridization and hybridization were carried out at 37°C for DNA-probes or 65°C for 2´OMe-probes. Membranes were washed three times for ≥ 1h in 2x SSC + 0,1% SDS and exposed to Phosphoimager Screens (Fujifilm; Tokyo, Japan) for up to 1 week. Screens were scanned using a Typhoon scanner (Amersham Biosciences) and band intensities were quantified using Multi Gauge software (Fujifilm; Tokyo, Japan). Membranes were immersed in 1% SDS heated to boiling in a conventional microwave and allowed to sit there for 5 min before they could be reused for prehybridization.

4.2.4.3 Analysis of mRNA levels by Polymerase Chain Reaction

4.2.4.3.1 Semi-quantitative method

1 µg total RNA extract was used for cDNA production with the Superscript II Reverse Transcriptase (Invitrogen; Karlsruhe, Germany). cDNA was diluted and used for PCR amplification with primers for the first dsRBD common to all *loqs* isoforms (Loqs_PB_RB1_BamHI_fw; Loqs_PB_RB1_BglII_rv). The following Phusion-polymerase PCR Mix was used:

4 µl Phusion HF buffer (5x)
0.4 µl 10 mM dNTP Mix
0.1 µl primer fw
0.1 µl primer rv
1 µl template cDNA
<u>0.2 µl Phusion Hot Start Polymerase</u>
20µl

PCR amplification was limited to 25 cycles and adapted for optimal conditions of the Phusion polymerase:

30 sec 98°C

24 cycles:
10 sec 98°C
30 sec 65°C
1 min 72°C

5 min 72°C
Hold 4°C

Reaction products were separated on a 1.5% agarose gel and band intensities were quantified with Multi Gauge software (Fujifilm; Tokyo, Japan).

4.2.4.3.2 Real-time quantitative PCR (RT-qPCR)

1 µg of total RNA extract was reverse transcribed according to the Qiagen miScript protocol:

4 µl miScript RT buffer (5x)
0.3-0.7 µl total RNA
14.3-14.7 µl H$_2$O
<u>1 µl miScript enzyme mix</u>
20 µl

Samples were incubated at 37°C for 60 min and then inactivated at 95°C for 5 min. After adding 100 µl of water to make a final volume of 120 µl, 1:10, 1:100 and 1:1000 dilutions *(post-RT dilutions)* were made. The qPCR reaction mix was as follows:

Reaction mixes for 14 reactions (for 1 row of 96-well plate):

70 µl Quantitect SyBr-green mix (2x)

35 µl H_2O

14 µl miScript universal primer (5 µM)

7 µl miScript specific primers (10 µM)

9 µl of reaction mix and 1 µl of RT-reaction per well was amplified in an ABI PRISM 7000 qPCR cycler (Applied Biosystems; Foster City, USA) using the following conditions:

15 min 94°C

40 cycles:
20 sec 94°C
30 sec 55°C
30 sec 70°C

Cycle of Threshold values (CT-values) usually determined via the auto-CT function and manually adjusted if necessary. The 2S-1 primer from the miScript kit was used as a control.

4.2.4.3.3 ΔΔ CT-Method

Expression levels were analyzed using the ΔΔ CT-Method (Schmittgen et al., 2008).

4.2.5 Drosophila melanogaster methods

4.2.5.1 Maintenance and handling

Flies were kept on standard fly food (see Materials 4.1.13) at 25°C and transferred to new food every 2-3 weeks. For phenotype selection flies were anesthetized with CO_2 and sorted on a CO_2-emitting pad (Genesee Scientific; San Diego, USA) using a Leica MZ7 stereomicroscope (Leica Microsystems; Wetzlar, Germany).

To slow proliferation by reducing metabolic rates flies were kept at 18°C for up to one week if necessary.

4.2.5.2 Transgenic flies

4.2.5.2.1 Obtaining transgenic lines

The myc-Loqs-PD and -PC constructs described elsewhere (see Materials 4.1.11.1; pEH37 and pEH38, respectively, see Results 5.17, Table 5) were sub-cloned into pUAST vector (Brand et al., 1993). The constructs, listed in Results 5.17 Table 3, were sequenced, tested in cell culture and sent for injection into w^{1118} fly embryos (Rainbow Transgenic Flies; Newbury Park, USA). Transgenic offspring, marked by red eye color, were twice crossed with w^{1118} flies to reduce the risk of secondary mutations. Siblings were then mated to produce homozygous stable lines (see Results 5.17 Table 4).

4.2.5.2.2 Mapping of P-element insertions by inverse PCR

The protocol for the mapping of P-element insertions was applied according to the Berkeley *Drosophila* Genome Project. In short, DNA was prepared from 30 anesthetized flies by freezing at -80°C and subsequent mechanical lysis in Buffer A. LiCl/KOAc-solution was added, debris was pelleted and supernatant was precipitated with isopropanol. The genomic fly DNA pellet was washed with 70% ethanol, dried and dissolved in H_2O. To create appropriate fragment sizes for inverse PCR, genomic DNA was digested with *HinP1*I or *Msp*I. The fragments were circularized by ligation at low DNA concentration. Standard PCR with Sp1 and Spep1 primers indicated above (see Materials 4.1.11.6.1) was carried out, amplification products were separated on a 1.5% agarose gel. Bands were excised, sequenced and mapped to the *Drosophila* genome. Insertion sites in genomic DNA samples of transgenic lines were verified by PCR with newly designed genomic primers flanking the putative P-element (see Materials 4.1.11.6.2).

4.2.5.2.3 Crossing

To obtain Loqs-PD (and Loqs-PC) rescue flies virgins of a tubulin-Gal4 driver line, Loqs-PB rescue flies and homozygous UAS-Loqs-PD (or Loqs-PC) transgenic flies carrying the P-element insertion on the third chromosome were each crossed to Kr/Cyo; D/TM6, Sb, Tb double balancer males, to obtain offspring with balanced 2nd (CyO) and 3rd (TM6, Tb, Sb) autosomes. F1 offspring from the Loqs-PB rescue flies was selected for D/TM6, Sb, Tb phenotypes to recover the loqs^{KO} background. F1 virgins from the tub-Gal4 cross and the UAS-Loqs-PD (or Loqs-PC) cross were both mated with balanced loqs^{KO} males and F2 siblings were crossed to obtain homozygous loqs^{KO} flies with a tub-Gal4/UAS-Loqs-PD (or Loqs-PC) phenotype.

4.2.6 Recombinant expression and purification of GST- or His$_6$-tagged Loqs isoforms

4.2.6.1 Recombinant expression

First attempts to recombinantly express Loqs isoforms in small scale were made by sub-cloning Loqs-PB in bacterial expression vectors pGex-6P-1 or pET-28a (see Materials 4.1.8 and Appendix 2; pEH 52). Loqs-PB was PCR amplified with Loqs_PB_RB1_BamHI_fw and Loqs_PB_RB3_NotI_rv, *BamHI/NotI* digested and then ligated into pGex. An analogous construct for pPET-Loqs-PA construct (lacking the last codon of the ORF before the stop codon) was made by Milijana Mirkovic-Hösle in our laboratory.

For recombinant expression the BL21 *E. coli* expression strain (carrying the extra plasmid pLysS encoding for lysozyme and a chloramphenicol resistance) was transformed. Briefly, 5 ml of a 50 ml primary overnight culture of BL21 cells growing in LB-medium + ampicillin (or kanamycin) + chloramphenicol + 1% glucose at 37°C were used to inoculate 100 ml LB-medium supplemented with appropriate antibiotics and culture was grown to an optical density OD$_{600}$ of 0.6. Expression was induced with 0.1 mM Isopropyl-β-D-thiogalactopyranosid (IPTG) and cells were harvested after 7h. Cell pellets were re-suspended in Lysis-buffer (+1% Triton X-100, + Complete® without EDTA, Roche Diagnostics) and frozen in liquid nitrogen. Samples were thawed on ice and sonified to reduce viscosity and improve solubility of proteins:

Sonifier settings:

Output control: 5-6

Duty cycle 20-30

1 min sonify, 30 sec on ice, 1 min sonify

Debris was pelleted and 50 µg of total protein extract were run on a 10% polyacrylamide gel. Gel was stained with colloidal Coomassie.

4.2.6.2 Affinity purification of recombinant proteins

250 µl Glutathione Sepharose 4 Fast Flow (GE Healthcare; Freiburg, Germany) was washed three times with wash buffer (1x PBS; 0.5% Saponin; 1x Complete® without EDTA), resuspended in 500 µl washing buffer and loaded into a spin column (MoBiTec; Göttingen, Germany). 5 µl of sonified and filtered protein extract was applied twice to the column (FT1 and 2). Column was washed three times with 5 ml wash buffer (wash1-3) and then eluted with glutathione elution buffer in 6 steps á 100 µl (E1-6). Equivalent amounts were separated on a 10% polyacrylamide gel. Gel was stained with colloidal Coomassie.

5 Results

5.1 A novel isoform of *loqs*

The observation that endo-siRNA mediated silencing depends on *loqs* in combination with *dcr-2* was a surprise because it suggested that a hybrid complex with components of the canonical miRNA and siRNA biogenesis pathways exists. However, this interpretation may have been an oversimplification because the *loqs* gene can give rise to at least three different mRNAs, each coding for a protein with distinct properties (Figure 6A; Forstemann et al., 2005; Jiang et al., 2005; Saito et al., 2005). Only one isoform, called Loqs-PB, is an essential partner of Dcr-1 in the biogenesis of miRNAs (Jiang et al., 2005; Park et al., 2007). Loqs-PA, which lacks 46 amino acids compared to Loqs-PB, also associates with Dcr-1 but does not suffice for miRNA silencing. A shift in the reading frame due to alternative splicing leads to a new stop codon in the Loqs-PC isoform. Instead of the third dsRBD the Loqs-PC protein is characterized by 54 specific amino acids. However, there was a possibility that not all isoforms of Loqs had been annotated yet. Therefore, I performed 3'-RACE experiments to detect novel splice variants from both genomic DNA and a genomic expression construct for Loqs-PC (Figure 6A). I recovered a new mRNA variant of *loqs* (*loqs*-RD), where an alternative polyadenylation in the third intron leads to a novel protein isoform that lacks the third dsRBD of Loqs-PA/-PB and contains 22 amino acids of new protein sequence (Figure 6A and B). Polyadenylation (poly-A) signals in *Drosophila* and humans are well conserved (Retelska et al., 2006) and consist of a defined polyadenylation signal located 10-30 bases upstream of the mRNA cleavage site and a less conserved U-rich sequence located within the first 30 nucleotides downstream of the cleavage site. Poly-A prediction yielded several possible poly-A motifs adequately spaced from the experimentally inferred site for Loqs-PD (see Appendix 5). Together with a T-rich stretch downstream in the genome (Figure 6B) this corroborated the experimental data. Sequencing of clones from genomic expression constructs confirmed the common poly-A signal for *loqs*-RA, -RB and -RC (Figure 6A). With Loqs-PB functioning in miRNA biogenesis, I hypothesized that one of the other Loqs isoforms might specifically act in endo-siRNA dependent silencing. In order to determine which of the four splice-variants of Loqs is involved in the endo-siRNA pathway, I needed a tool to manipulate them individually.

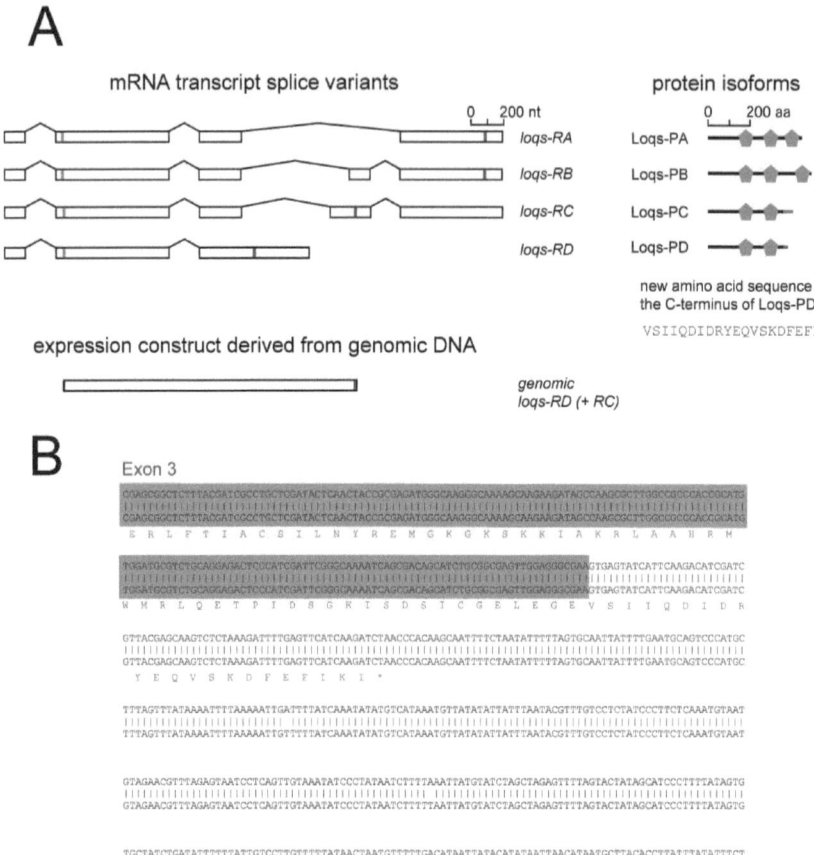

Figure 6: A novel isoform of the dsRBP Loquacious
A) Schematic diagram of the four *loqs* mRNA and protein variants currently known. In the mRNA exons are represented by horizontal bars, introns by thin lines. Start codons are indicated in green, stop codons in red. The dsRBD-motifs are depicted in the cartoon drawings of the protein isoforms (diagram analogous to Forstemann et al., 2005). The novel amino acid sequence of Loqs-PD is indicated. The genomic expression construct encodes for both *loqs*-RD and -RC, but expresses almost exclusively Loqs-PD (see **Figure 7C**)
B) 3´-RACE PCR for detection of the *loqs*-RD isoform; top strand of DNA represents *Drosophila melanogaster* genomic sequence, bottom strand represents cloned 3´-RACE product; the appended poly-A tail is colored in green; the amino acid sequence of Loqs-PD is indicated in red, the exon sequence common to all isoforms is shaded in blue

5.2 Isoform-specific knock-down

Isoform-specific knock-downs are possible if the corresponding mRNA contains a stretch of unique sequence that is amenable to RNAi. In the case of the *loqs* gene, it is possible to target the *loqs*-RC and *loqs*-RD RNA individually, the *loqs*-RB RNA together with the *loqs*-RC RNA, *loqs*-RA, *loqs*-RB and *loqs*-RC together via the common sequence towards the 3'-end and finally all four *loqs* isoforms simultaneously with dsRNA directed against the amino-terminus of Loqs (Figure 7A).

Detection of endogenous Loqs protein isoforms on immunoblots has revealed three bands of distinct sizes, which had been assumed to correspond to Loqs-PB, Loqs-PA and Loqs-PC (Forstemann et al., 2005). With the help of Loqs-PD-specific RNAi I could show that the smallest of these bands predominantly contains the new Loqs-PD isoform (Figure 7B). Difficulties to recover *loqs*-RC clones in my 3'-RACE experiments indicated that *loqs*-RC mRNA is only detectable in low levels in S2 cells. Therefore I tested whether it is expressed at all from the genomic construct encoding for both Loqs-PC and the novel Loqs-PD isoform. The genomic Loqs-PC/-PD expression construct produces two protein bands previously assumed to represent myc-tagged Loqs-PC and untagged Loqs-PC, translated from the endogenous start codon situated downstream of the myc-sequence. When cells were simultaneously transfected with the genomic expression construct and RNAi triggers, no protein bands could be detected after *loqs*-RD knock-down (Figure 7C). A control dsRNA, on the other hand, did not affect expression, indicating that the genomic construct (henceforth abbreviated Loqs-PD$_{genomic}$) almost exclusively expresses Loqs-PD. This indicates that the original detection of Loqs-PC might be a cloning artifact and that the isoform plays only a minor role *in vivo*. However, for reasons of completeness I included Loqs-PC in subsequent experiments. Immunoblotting confirmed that on protein level, the Loqs isoform-specific RNAi was efficient and specific (Figure 7B). The *loqs*-ORF dsRNA was slightly more successful in Loqs-PD depletion than the *loqs*-5'UTR dsRNA, while neither Loqs-PB nor -PA bands remained detectable. Alternative polyadenylation of the Loqs-PD isoform was substantiated by exclusive targeting of Loqs-PB, -PA (and -PC) by the dsRNA derived from the 3'UTR of these isoforms.

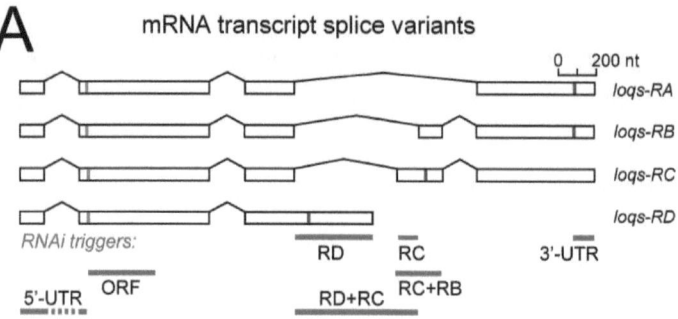

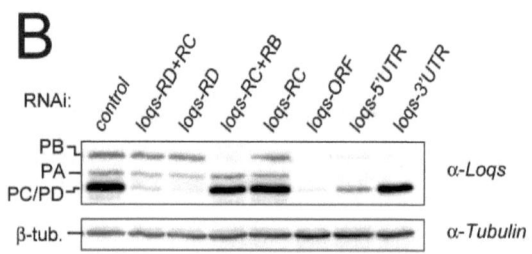

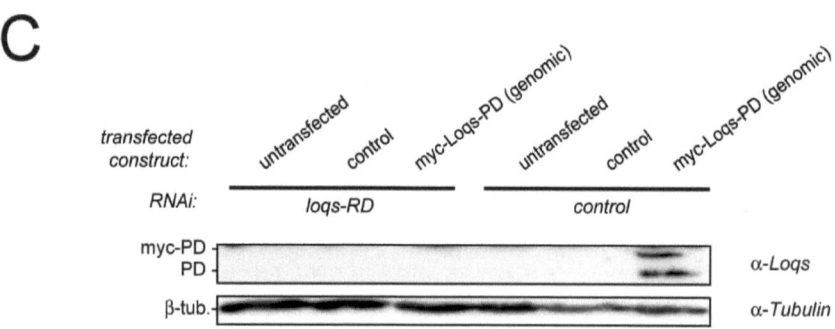

Figure 7: Loqs isoform-specific RNAi and verification on protein level
A) Schematic diagram of the four *loqs* mRNA variants; the regions from which dsRNA was derived to trigger isoform-specific RNAi are indicated below the mRNA variants; note that *loqs*-RD does not share the same 3´UTR with *loqs*-RA/-RB/-RC
B) Effect of isoform-specific RNAi treatment in *Drosophila* S2 cells; dsRNA against dsRed served as a control; the Western blot was probed with α-Loqs monoclonal antibody to confirm depletion of specific Loqs protein isoforms; α-tubulin served as a loading control (legend continued on p. 64)

(legend Figure 7 continued)
C) The genomic expression construct for Loqs-PD almost exclusively expresses Loqs-PD and not Loqs-PC; 5d after RNAi treatment against Loqs-PD or a dsRed control, S2 cells were split 1:5 and transfected with the genomic Loqs-PD expression construct or a pUC18 control; protein extract was prepared 4d after transfection; α-Loqs Western detected expression from the genomic construct only after control RNAi (Note that expression of endogenous Loqs isoforms is below the detection limit in this Western blot so that endogenous Loqs-PC expression cannot completely be ruled out); α-tubulin Western served as a control

In addition to the protein level of individual isoforms I tested for depletion of the corresponding mRNA by isoform-specific PCR. After isoform-specific RNAi treatment total RNA was extracted from S2 cells and cDNA was produced. With a primer pair spanning from the third to the last exon (indicated by arrows in

Figure 8A) I could distinguish three bands amplified from cDNA of *loqs*-RA, -RB and -RC isoforms and a minor contamination with genomic DNA (

Figure 8B). Note that the shortest protein isoform, Loqs-PC, corresponds to the largest band in the mRNA assay due to the extended fourth exon (

Figure 8A). Targeting by the RNAi triggers proved to be equally efficient and selective on RNA as on protein level. The *loqs*-RD cDNA was similarly detected with a primer pair within the *loqs*-RD-specific sequence (

Figure 8A, C; amplicon indicated by green bar). Treatment of S2 cells with the RNAi trigger directed against the 3´UTR of Loqs-PA, -PB and -PC did not affect cDNA levels of *loqs*-RD while targeting the second exon, common to all splice variants (*loqs*-ORF), depleted the PD-specific amplicon. However, the PCR analysis did not allow a quantification of overall Loqs levels. It was possible that depletion of one transcript variant would relatively enrich the other ones, making interpretation of isoform-specific RNAi data difficult. To test this I used both a semi-quantitative and a quantitative PCR (qPCR) approach to determine total *loqs* cDNA levels prepared from S2 cells after isoform-specific RNAi (

Figure 8D). For the semiquantitative method equal amounts of cDNA were used as a template for PCR amplification with a primer pair flanking the first dsRBD common to all isoforms (Loqs_PB_RB1_BamHI_fw; Loqs_PB_RB1_BglII_rv; see Materials and Methods 4.1.11.1.1). Amplification was stopped within the exponential phase, products were separated on an agarose gel and band intensities were quantified. Results were comparable to SyberGreen-based qPCR with a specially designed primer pair: *loqs*-RC depletion, consistent with its low expression level, caused no reduction of total *loqs* levels compared to an RNAi control (

Figure 8D). If isoform-specific RNAi is not compensated by increased transcription of the re-

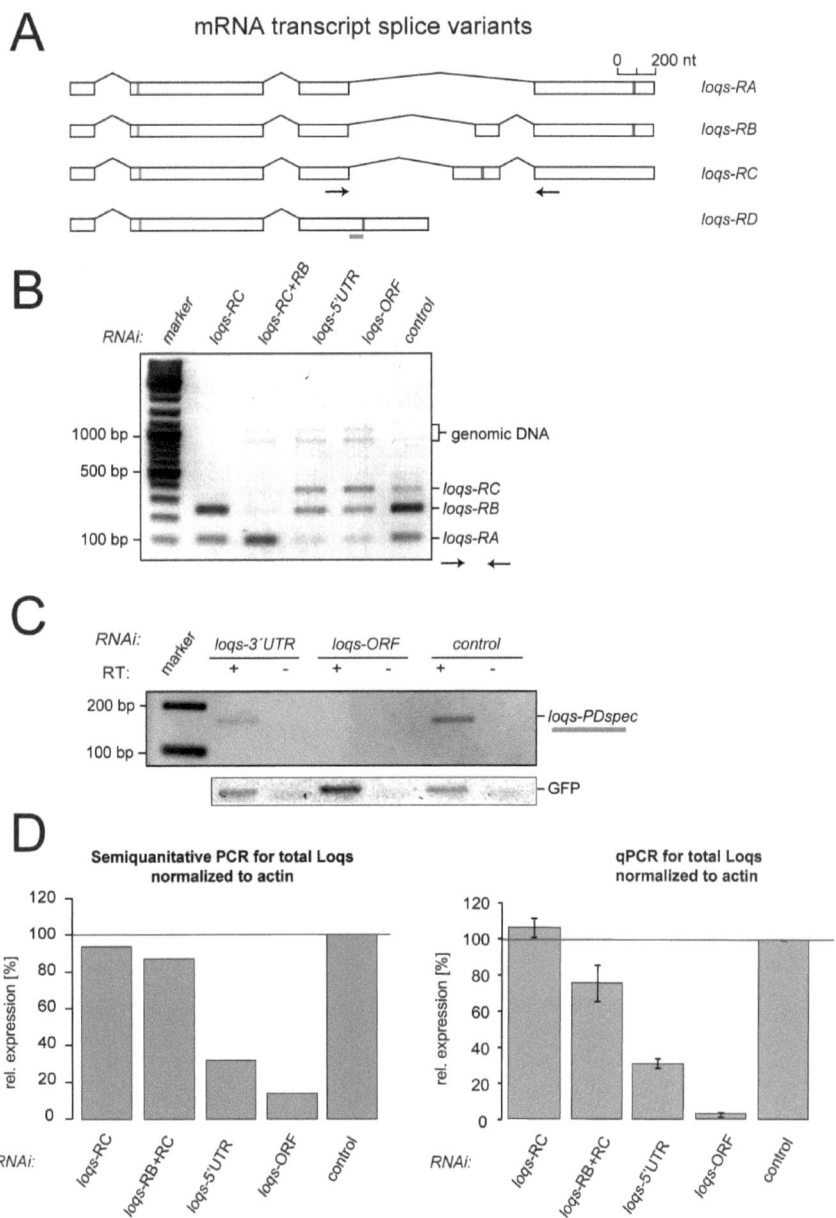

Figure 8: Loqs isoform-specific RNAi and verification on RNA level (legend continued on p. 66)

(legend Figure 8 continued)
A) Schematic diagram of the four *loqs* mRNA variants; arrows indicate the primer pair used for amplification of *loqs*-RA/-RB/-RC from cDNA; the PD-specific amplicon is marked by a green horizontal bar below the *loqs*-RD sequence
B) Effect of isoform-specific RNAi treatment on mRNA levels; total RNA of S2 cells was extracted and cDNA was amplified with the primer pair for *loqs*-RA/-RB/-RC detection (compare arrows in **A**); PCR products were separated on a 1.5% agarose gel and stained with SyberSafe; main bands of a 1 kb DNA marker mix are indicated
C) Effect of isoform-specific RNAi treatment on mRNA level of *loqs*-RD; total RNA of S2 cells was extracted; samples were either reverse transcribed (RT +) or not (RT -) and then amplified with the primer pair for *loqs*-RD detection (compare green bar in **A**); *loqs*-3´UTR RNAi did not deplete the *loqs*-RD mRNA; PCR products were separated on a 4% agarose gel and stained with SyberSafe; main bands of a 1 kb DNA marker mix are indicated; amplification of a ca. 100 bp GFP fragment served as a control for cDNA quality
D) Effect of isoform-specific RNAi on total *loqs* RNA level;
Left panel: semi-quantitative amplification of cDNA from S2 cells with the L1_fw/L1_rv primer pair for all four *loqs*-isoforms; after 25 PCR cycles products were separated on an agarose gel, stained with SyberSafe and band intensities were quantified with MultiGauge software (Fujifilm) relative to an actin control; RNAi triggers are indicated below the bars; horizontal red line indicates no change compared to a dsRed RNAi control
Right panel: qPCR from total cDNA preparations of S2 cells; qPCR primers for total *loqs* were used; expression was normalized to a 2S control; horizontal red line indicates no change compared to a dsRed RNAi control; values are represented as mean ± SD (n=3).

maining isoforms, then additional RNAi against Loqs-PB should lead to more pronounced depletion of total *loqs*. Indeed, qPCR data showed significantly reduced *loqs* expression after *loqs*-RB+RC treatment (

Figure 8D, second panel). Both methods demonstrated that, consistent with findings on protein level, more total *loqs* cDNA remained after treatment with the 5´UTR trigger than with the *loqs*-ORF dsRNA (

Figure 8D).

Together, my isoform-specific RNAi constructs are selective and deplete targeted mRNAs and corresponding proteins effectively, without interfering with the expression of remaining isoforms.

5.3 A cell culture reporter system for endo-siRNA silencing activity

In order to study the endo-siRNA pathway in cell culture with an easily visible and quantifiable readout I re-examined the GFP reporter cell lines in our laboratory stock. The reasoning was that some of the canonical features of transposable elements, such as multicopy insertion and the formation of repetitive regions, are shared by transgenes that have integrated into the host cell genome after transfection and selection of stable cell culture lines. Indeed, Ago2-dependent repression of a stably integrated GFP expression plasmid in *Drosophila* cells has been described (Saito et al., 2005).

Several GFP-based reporter cell lines were derived from S2 cells transfected with a plasmid (pKF63; see Appendix 3) encoding an Enhanced Green Fluorescent Protein (EGFP) transgene under the control of a two kilobase ubiquitin promotor and a downstream SV40 polyadenylation signal (

Figure 9A). The additional ampicillin resistance gene allows for bacterial selection. The pKF63 plasmid was co-transfected with an antibiotic resistance vector (pHSneo for G418 neomycin resistance) to allow selection of stable, green fluorescent cells. Stable cells were diluted to isolate clonal lines, each derived from a single cell (

Figure 9B). In our cell culture stock we had two clonal lines prepared by Klaus Förstemann with comparatively low (63-6) and intermediate (63N1) levels of GFP fluorescence. The hypothesis was that in both lines endo-siRNAs from the high-copy GFP transgene would cause a certain degree of silencing of GFP mRNAs. This would lead to visible changes in fluorescence levels when the efficacy of endo-siRNA silencing changed.

A

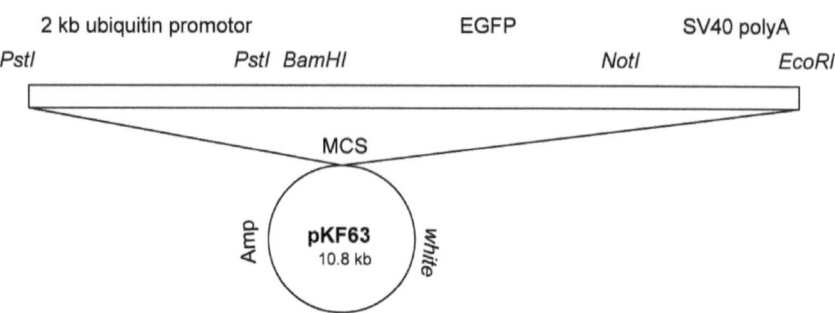

B

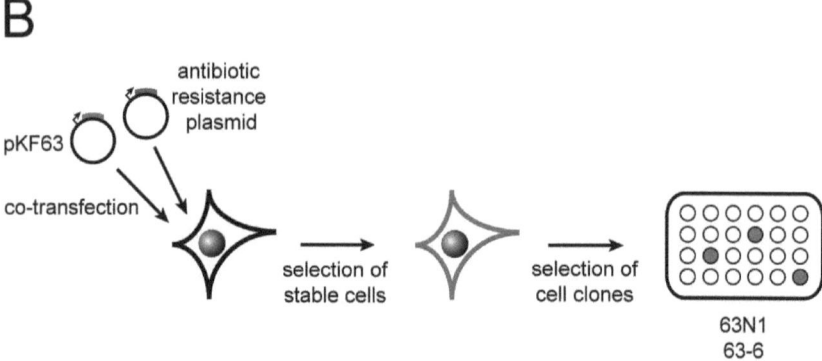

Figure 9: The endo-siRNA cell culture reporter
A) 10.8 kb pKF63 expression construct (Forstemann et al., 2005); a 2 kb ubiquitin promoter drives expression of enhanced GFP (EGFP); the SV40 poly-A signal was added downstream of the GFP into the Multiple Cloning Site (MCS); ampicillin resistance gene for bacterial selection; restriction enzymes used for cloning are indicated
B) The endo-siRNA reporter system is based on transfection of a GFP expression construct together with an antibiotic resistance plasmid (pHSneo) to allow for selection of stable transformants, followed by selection of stable, single-cell derived clones; the two cell lines analyzed in this thesis are indicated below the cartoon drawing of the 96-well plate

I tested whether the 63N1 cell line could serve as a cell culture reporter for endo-siRNA silencing by measuring the GFP levels after RNAi against small RNA silencing components (

Figure 10A). From the three RNaseIII-enzymes Drosha, Dcr-1 and Dcr-2, only the latter appeared to be involved in repression of GFP. Depletion of the cytoplasmic dsRBD-protein Loqs also resulted in a de-repression of GFP, while depletion of its homolog R2D2 appeared to increase repression. Finally, Ago2 is the main effector protein mediating this response, although depletion of Ago1 also resulted in a slight de-repression (Figure 7A). These are precisely the genetic requirements of the endo-siRNA pathway in *Drosophila* (Czech et al., 2008; Kawamura et al., 2008; Okamura et al., 2008b). With the endo-siRNA cell culture reporter system and the newly established isoform-specific RNAi I had the tools to determine the roles of individual Loqs isoforms in endo-siRNA dependent silencing.

5.4 Loqs-PD is essential for silencing high-copy transgenes

Employing isoform-specific knock-downs in the 63N1 cell line I saw that GFP fluorescence did not change upon depletion of *loqs*-RC alone, while targeting *loqs*-RA+RB+RC (compare *loqs*-3´UTR) even caused hyper-repression of the reporter (

Figure 10B). In contrast, specific targeting of *loqs*-RD as well as targeting *loqs*-RD and *loqs*-RC together led to an even stronger de-repression than knock-down of all *loqs* variants. Thus, Loqs-PD appears to be required for repetitive-element-derived endo-siRNA silencing in our artificial reporter.

If Loqs-PD is not only necessary but sufficient for endo-siRNA silencing, it should be able to rescue the effect of RNAi against all endogenous isoforms. Endo-siRNA reporter cells were treated with *loqs*-5´UTR dsRNA and transfected with expression constructs for myc-tagged Loqs-PB, -PA or the genomic Loqs-PD construct, none of which is targeted by the RNAi trigger against endogenous *loqs*. Only Loqs-PD expression could revert the de-repression of the reporter caused by RNAi (

Figure 10C). However, Western blotting showed only a 20-35% reduction of endogenous Loqs so that over-expression alone could account for the observed effect on the endo-siRNA cell culture reporter (compare below).

To optimize the readout of the reporter I roughly determined the kinetics of RNAi against *loqs* isoforms (

Figure 10D). The amplitude of the reporter reaction is dependent on the amount of Loqs-PD protein. De-repressive effect of *loqs-ORF* and hyper-repressive effect of *loqs-RB+RC* treatment increased continually during the five day assay. All RNAi experiments were therefore assayed after a 5-6 day interval.

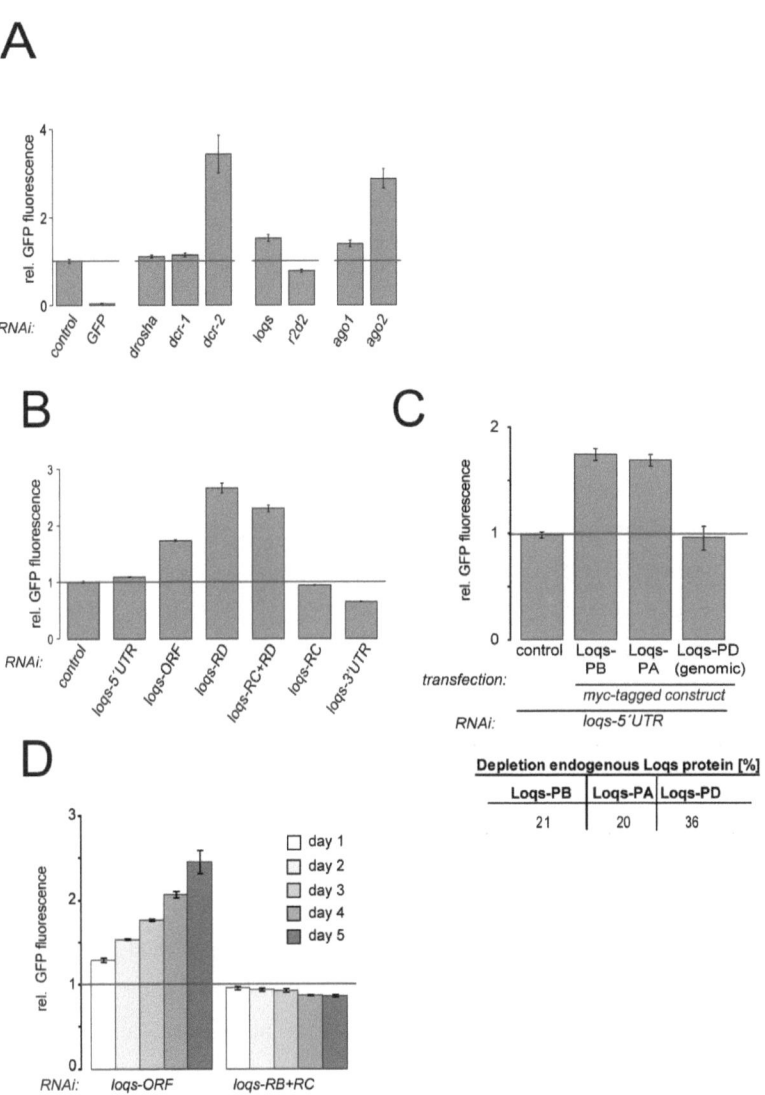

Figure 10: Loqs-PD is essential for endo-siRNA dependent silencing of stably integrated transgenes (legend continued on p. 71)

(legend Figure 10 continued)
A) Individual components of small RNA silencing pathways were depleted via RNA interference to define the genetic requirements for GFP repression in the 63N1 cell culture reporter; RNAi triggers are indicated below the bars; the control dsRNA was directed against DsRed; values are represented as mean ± SD (n=3); the horizontal red line marks no change compared to the control

B) Effects of isoform-specific RNAi experiments on the GFP-expression in the 63N1 cell culture reporter; RNAi triggers are indicated below the bars; values represent the mean ± SD (n=3); the horizontal red line marks no change compared to the control

C) Loqs-PD is sufficient to rescue knock-down of endogenous Loqs; the endo-siRNA reporter cell line was treated with RNAi against the 5´UTR of all endogenous *loqs* isoforms and transfected with plasmids for expression of myc-tagged Loqs-PA, -PB and the genomic expression construct for myc-Loqs-PD; values represent the mean ± SD (n=3); the horizontal red line marks no change compared to the mock-transfected control; table shows efficiency of endogenous Loqs depletion on protein level: Western blots from protein extracts were probed with α-Loqs antibody and band intensities of individual isoforms were quantified using MultiGauge software (Fujifilm) and normalized to a tubulin loading control. Depletion efficiency of endogenous Loqs isoforms was determined by comparing the *loqs*-5´UTR sample to a mock control.

D) Kinetics of endo-siRNA reporter reaction after isoform-specific RNAi treatment; RNAi triggers are indicated below the bars; the horizontal red line marks no change compared to dsRed control; a single experiment is shown

5.5 Loqs-PD is essential for biogenesis of hairpin-derived endo-siRNAs

In order to test whether Loqs-PD is required for naturally occurring endo-siRNAs as well, I depleted individual Loqs protein isoforms in S2 cells and then measured the levels of an endo-siRNA derived from the long hairpin-forming gene *CG4068* by Northern blotting (Czech et al., 2008; Kawamura et al., 2008; Okamura et al., 2008b). Only depletion of the *loqs*-RD transcript correlated with a reduced production of the *CG4068* B endo-siRNA (Figure 11A), while the biogenesis of the *bantam* miRNA was not impaired when *loqs*-RD was targeted. Depletion of all other isoforms (*loqs*-3´UTR) might even enhance endo-siRNA biogenesis (Figure 11A, B). Knock-down of another factor associated with endo-siRNA silencing, Dcr-2, similarly reduced biogenesis of the *CG4068* B endo-siRNA, indicating that Dcr-2 and Loqs-PD are equally important in the process (Figure 11B). As expected, depletion of the Loqs-PB isoform (compare *loqs*-5´UTR, *loqs*-ORF and *loqs*-3´UTR) led to accumulation of pre-*bantam*, indicating impaired miRNA biogenesis (Figure 11A). Similarly, depletion of Dcr-1, the RNAseIII enzyme for miRNA biogenesis, affected biogenesis of both *bantam* and miR-277 (Figure 11B). *CG4068* B levels, on the other hand, appeared increased, comparable to *loqs*-3´UTR treatment. Ago1 depletion only decreased the stability of mature *bantam*, since both endo-siRNAs and miR-277 are preferentially loaded into Ago2 RISC complexes (Figure 11B; Förstemann, 2007; Czech et al., 2008; Kawamura et al., 2008). Northern blot data were verified by qPCR with primers for *CG4068* (Figure 11C). While both total Loqs and Dcr-2 depletion reduced *CG4068* expression levels by more than 50%, *loqs*-3´UTR RNAi indeed caused higher expression levels of the hairpin-derived endo-siRNA. This suggests that efficiency of endo-siRNA silencing is directly correlated with the level of Loqs-PD, while other isoforms may have an inhibitory effect.

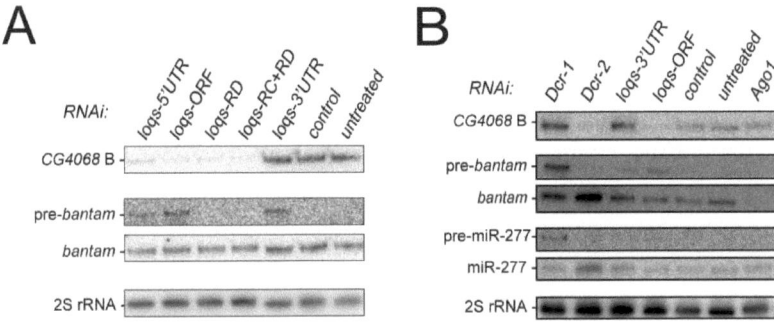

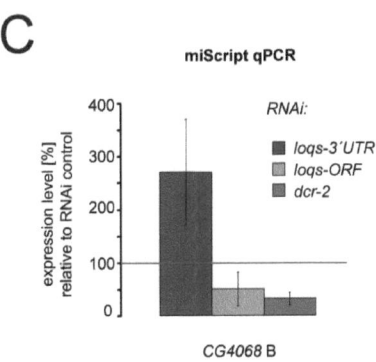

Figure 11: Loqs-PD is essential for biogenesis of hairpin-derived endo-siRNAs
A) Northern blot after Loqs isoform-specific RNAi in *Drosophila* S2 cells; RNAi triggers are indicated; dsRed RNAi served as a control; probes against the hairpin-derived endo-siRNA (called *CG4068* B in Okamura et al., 2008b) and the *bantam* miRNA were used; pre-*bantam* marks the accumulation of un-diced precursor; 2S rRNA served as a loading control
B) Northern blot after RNAi against components of small RNA silencing pathways in *Drosophila* S2 cells; RNAi triggers are indicated; dsRed RNAi served as a control; probes against the *CG4068* B as well as *bantam* and miR-277 miRNAs were used; pre-miRNA marks the accumulation of un-diced precursor; 2S rRNA served as a loading control
C) qPCR analysis of *CG4968* B expression after depletion of small RNA silencing components; qPCR primers for *CG4068* B were used; expression is normalized to a 2S-1 control; horizontal red line indicates no change compared to a dsRed RNAi control; values are represented as mean ± SD (n=3)

5.6 Loqs-PD-specific antibody

The Loqs-PD-specific sequence not only makes the *loqs-RD* mRNA amenable to isoform-specific RNAi, but can act as an epitope for antibody recognition. I ordered immunization of rabbits with the 22 amino acid peptide that is unique to the Loqs-PD isoform. The specificity of the affinity-purified antibodies was confirmed in Western blotting after isoform-specific RNAi of endogenous *loqs* (Figure 12A, first panel). The antibody also detected overexpressed and recombinant Loqs-PD (Figure 12A, second panel) and could recognize the PD-specific domain in a fusion protein with GFP (Figure 12B). Results were comparable for antibody samples from two different rabbits (data not shown) and both samples could be used for co-immunoprecipitation experiments as well.

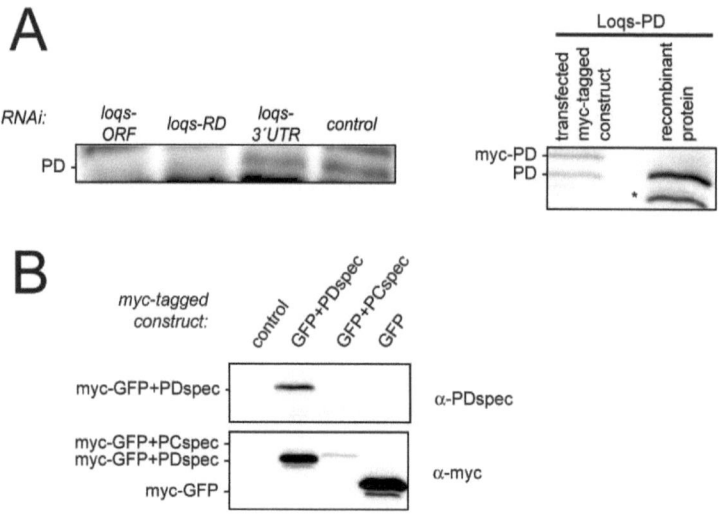

Figure 12: Test of affinity-purified PD-specific antibody for specificity
A) The PD-specific antibody detects overexpressed and endogenous Loqs-PD
Left panel: Western blot from S2 cell extract after isoform-specific RNAi; RNAi against dsRed served as a control
Right panel: Western blot from S2 cell extract after transfection with a myc-tagged *loqs-PD* expression construct as well as purified recombinant Loqs-PD protein; the asterisk marks protein degradation
B) Western blot from S2 cells transfected with myc-tagged GFP-fusion constructs; Loqs-PD-specific antibody can detect GFP-PDspec chimaeras; α-myc Western served as a control for expression

5.7 Loqs-PD interacts with Dcr-2 in cell culture and flies

In order for Loqs-PD to act in the biogenesis of endo-siRNAs, it must partner with Dcr-2. Co-immunoprecipitation of Loqs with Dcr-2 has been described previously (Czech et al., 2008) but the antibody employed recognized all Loqs protein isoforms. With the help of cDNA constructs coding for only one of the splice variants, I could distinguish Loqs-PA, -PB and -PD in transfected S2-cells and determine the potential of each isoform to interact with Dcr-2 by co-immunoprecipitation. Epitope-tagged Loqs-PD, just like tagged R2D2, was able to associate with Dcr-2 but not with Dcr-1, though the extent of Dcr-2 association varied between experiments (

Figure 13A-C). This is consistent with the previous observation that on the level of the endogenous protein, the smallest Loqs isoform (identified as Loqs-PD in this thesis) does not co-immunoprecipitate with Dcr-1 (Forstemann et al., 2005). In contrast, tagged Loqs-PB and Loqs-PA can associate with Dcr-1, but also to some extent with Dcr-2. Since only Loqs-PD is required for endo-siRNA generation, it appears that the Dcr-2/Loqs-PD complex is exclusively responsible for endo-siRNA generation.

To exclude the possibility that Loqs-PD/Dcr-2 binding is due to an overexpression artifact, I used the PD-specific antibody to precipitate endogenous Loqs-PD in untreated S2 cells. This resulted in significant co-precipitation of endogenous Dcr-2, indicating strong interaction of Loqs-PD and Dcr-2 at endogenous levels (

Figure 13D).

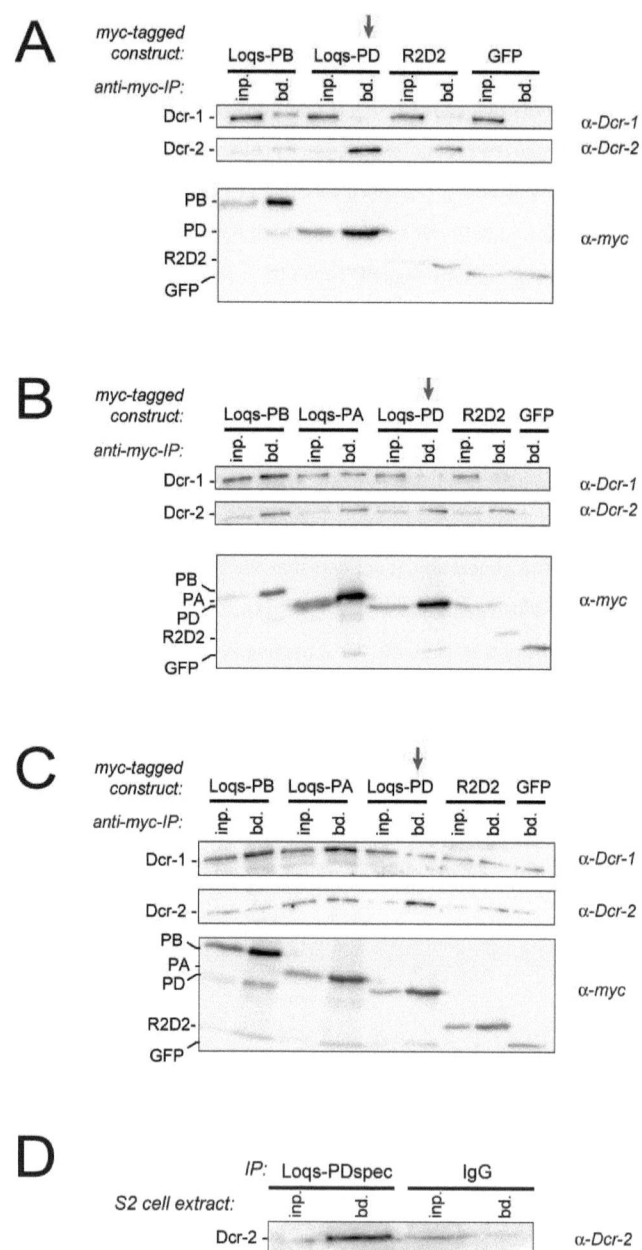

Figure 13: Dcr-2 preferentially interacts with Loqs-PD (legend continued on p. 76)

(legend Figure 13 continued)

A-C) α-myc co-immunoprecipitation of endogenous Dcr-1 and Dcr-2 from S2 cells expressing myc-tagged Loqs isoforms and R2D2. Three independent experiments are shown. Although the extent of association varied, Loqs-PD associates preferentially with Dcr-2 (compare red arrow). While some association between Loqs-PA/-PB and Dcr-2 may also occur, these two isoforms preferentially associate with Dcr-1; R2D2 served as a positive control for Dcr-2 interaction; α-myc Western served as a control for co-immunoprecipitation; inp. = input; bd. = bound

D) Co-immunoprecipitation of endogenous Dcr-2 with endogenous Loqs-PD from *Drosophila* S2 cell extract; α-Loqs-PD specific antibody was used for IP; rabbit immunoglobulin G (IgG) served as a control; inp. = input; bd. = bound

5.8 The PD-specific C-terminus is sufficient for Dcr-2 binding and essential for endo-siRNA function

Since Loqs-PB binding to Dcr-1 is mediated via the C-terminus containing the third dsRBD (Forstemann et al., 2005; Ye et al., 2007), I wondered whether the Loqs-PD-specific C-terminus is sufficient to interact with Dcr-2. I fused the unique sequences of Loqs-PD as well as Loqs-PC to GFP and performed anti-GFP immunoprecipitations. A small amount of Dcr-2 co-precipitated reproducibly in case of the GFP-PDspec fusion (Figure 14A), demonstrating that the PD-specific part can be sufficient for interaction with Dcr-2. Nonetheless, this interaction is weaker than the one observed with full-length Loqs-PD (

Figure 13).

Since the PD-specific sequence was sufficient for Dcr-2 interaction, I tested whether it was required for endo-siRNA dependent silencing. I transfected the endo-siRNA cell culture reporter with *loqs-PA*, *-PB* and *-PC* expression vectors. Overexpression of the isoforms not known to be involved in endo-siRNA silencing caused an increase in GFP expression (Figure 14B). Transfection with a version of *loqs* truncated at the start of the PD-specific amino acids resulted in comparable impairment of endo-siRNA silencing, suggesting that this protein is non-functional in the endo-siRNA pathway (Figure 14B). Re-addition of the Loqs-PD-specific sequence to the truncated Loqs variant to reconstitute Loqs-PD (L1L2+PDspec containing two ligation-dependent extra amino acids, glycine and lysine) completely reverted the dominant-negative effect (Figure 14B). Taken together, my results indicate that the different isoforms of Loqs can compete for binding to Dcr-2 and that the C-terminus of Loqs-PD is required for endo-siRNA silencing.

Initial studies analyzing the genetic requirements for endo-siRNA silencing found no necessity for *r2d2*, a paralog of *loqs* (Czech et al., 2008; Zhou et al., 2009). R2D2 contains two dsRBDs and interacts with Dcr-2 via its C-terminal part (Liu et al., 2003; Ye et al., 2007). I exploited the similar domain architecture of Loqs and R2D2 by swapping either the PD-specific sequence (PDspec) or the R2D2 C-terminus (RC-term) between the Loqs and R2D2 scaffolds consisting of the first two dsRBDs (L1L2 and R1R2, respectively; compare schematic drawing in Figure 14C). Overexpression of the wild-type proteins had no effect (R2D2) or caused slight hyper-repression (Loqs-PD) of the endo-siRNA reporter. In contrast, expression of an R1R2+PDspec chimaera resulted in increased GFP expression, indicating that endo-siRNA mediated silencing had been impaired (Figure 14C, black bars). Apparently, the R1R2+PDspec construct is a dominant negative protein that may be able to sequester endo-siRNA pathway components such as Dcr-2 but cannot functionally substitute Loqs-PD. The truncated form of Loqs (L1L2) led to a two-fold de-repression of the endo-siRNA reporter (Figure 14C, black bars). Interestingly, an L1L2+RC-term chimaera impaired endo-siRNA silencing to a lesser extent, indicating that the Dcr-2 binding element from R2D2 can partially substitute the Loqs-PD-specific sequence.

I observed a distinct reaction to the overexpressed proteins with a miRNA reporter system, where the GFP mRNA contains two perfect binding sites for miR-277 (Figure 14C, white bars). This reporter is silenced by Ago2-loaded miR-277, a process which depends on the Dcr-2/R2D2 RISC loading complex (Förstemann, 2007). In these cells the R1R2+PDspec construct is functional, presumably because the PD-specific sequence can substitute the native C-terminus and mediate association with Dcr-2. On the other hand, Loqs-PD overexpression impairs miR-277 dependent GFP-silencing, indicating that Loqs-PD cannot substitute R2D2 in the miR-277 programmed RISC loading complex (Figure 14C, white bars).

I tried to express myc-tagged Loqs-PD-specific sequence alone for potential interaction studies and reporter assays. However, neither Western blotting nor dot blotting allowed detection of the isolated PD-specific sequence (Figure 14D). Since the sequencing data for the expression construct verified correct cloning, this may indicate degradation of the peptide.

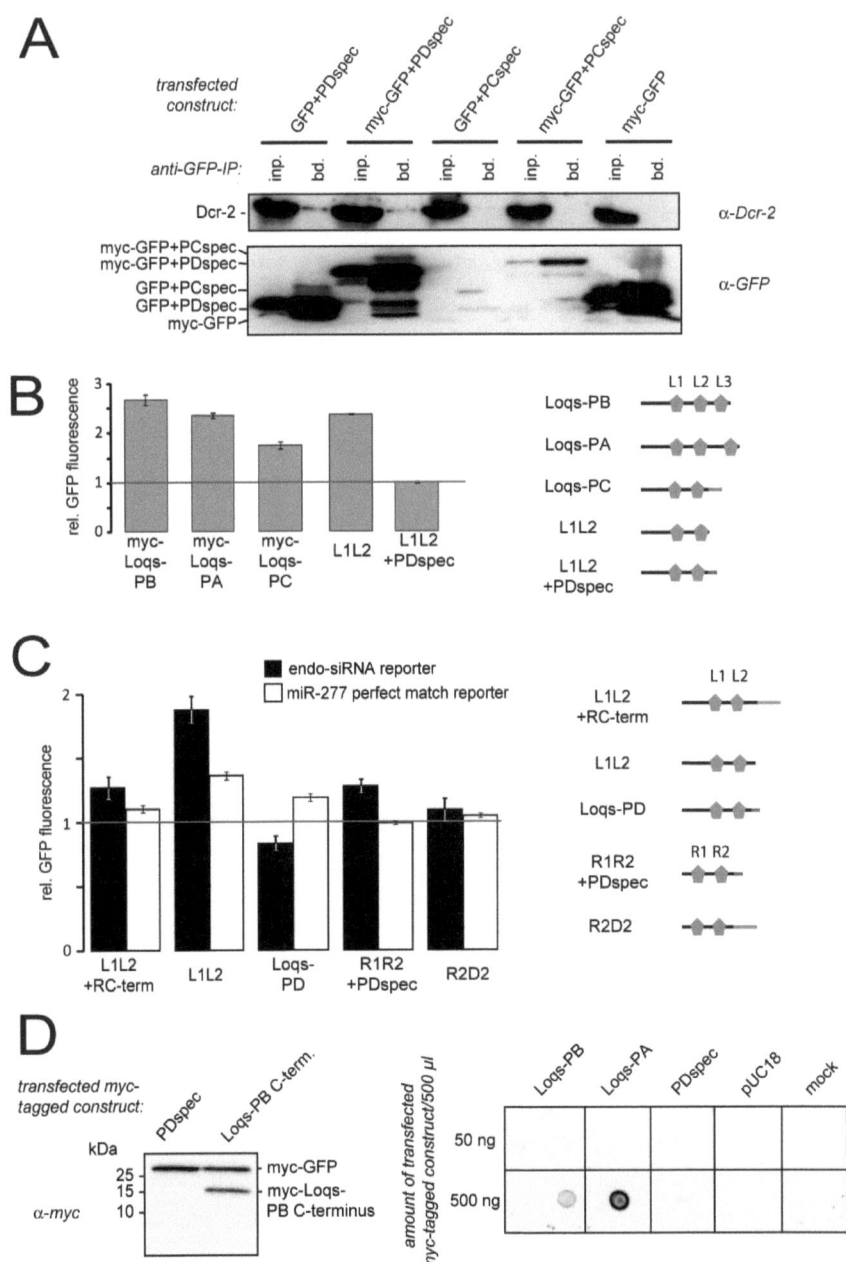

Figure 14: Dcr-2 interacts with the C-terminal 22 amino acids specific to Loqs-PD (legend continued on p. 79)

(legend Figure 14 continued)
A) Western blot after co-immunoprecipitation of endogenous Dcr-2 with GFP-fusion constructs in S2-cells; blot was stripped and re-probed for α-GFP to confirm expression and immunoprecipitation; inp. = input; bd. = bound
B) Effect of Loqs isoform overexpression on GFP expression levels of the endo-siRNA cell culture reporter (Loqs-PA, -PB and -PC contained an N-terminal myc-tag); protein isoforms are depicted on the right (dsRBDs are represented by brown symbols; Loqs-PC-specific sequence is colored in blue, Loqs-PD-specific sequence in red); measurement values represent the mean ± SD (n=3) and were normalized to a pUC18 transfected control (red line)
C) Effect of third domain swaps between Loqs (L1L2) and R2D2 (R1R2) scaffolds on GFP expression levels of the endo-siRNA cell culture reporter (black bars) and the miR-277 perfect match reporter (white bars; Förstemann, 2007); protein isoforms are depicted on the right (dsRBDs are represented by brown symbols for Loqs, green symbols for R2D2; Loqs-PD-specific sequence is colored in red, the R2D2 C-terminus is colored in green); measurement values represent the mean ± SD (n=3) and were normalized to a pUC18 transfected control (red line)
D) Test for expression of myc-tagged PD-specific sequence;
Left panel: α-myc Western blot of 63N1 cell extract transfected with expression constructs for myc-tagged PD-specific sequence or Loqs-PD C-terminus; no band for the PD-specific peptide could be detected; endogenous myc-GFP from 63N1 cells served as a loading control; molecular weight marker indicated on the left
Right panel: Dot blot for expression of myc-PD-spec; S2 cells were transfected with 50 ng or 500 ng of myc-tagged Loqs-PA/-PB or PD-spec peptide; pUC18 and mock transfection served as controls; protein extracts were spotted on a nitrocellulose membrane and probed for expression with α-myc antibody

5.9 Loqs-PD interacts with the N-terminal helicase domain of Dcr-2 during endo-siRNA biogenesis

My results suggest that the 22 amino acids unique to Loqs-PD convey its binding capacity for Dcr-2 and an essential part of its functionality during endo-siRNA silencing. I therefore proceeded to characterize the corresponding interaction domain in Dcr-2. Dcr-2 contains two N-terminal helicase domains, a Domain of Unknown Function (DUF) and a central PAZ domain. Two endonucleolytic RNAseIII domains and a dsRBD are situated at the C-terminus (Figure 15A). Recently, an EM-model proposed an L-shaped arrangement for human Dicer with the catalytic domain residing in the long branch (Lau et al., 2009; Wang et al., 2009). The N-terminal DExH/D helicase domain of human Dicer, situated in the short branch of the structure, mediates binding to the mammalian Loqs homologue TRBP (Haase et al., 2005; Lau et al., 2009; Wang et al., 2009). I therefore tested if *Drosophila* Dcr-2 also interacts through its N-terminal helicase domain with Loqs-PD by co-expressing Flag-tagged Dcr-2 with GFP+PDspec/+PCspec fusions in S2 cells. Full-length Dcr-2 efficiently co-precipitated with the GFP+PDspec fusion proteins, corroborating the observations for endogenous Dcr-2 (Figure 15B). When I substituted the full-length Flag-Dcr-2 with a construct lacking the N-terminal helicase region (Δ1-551 = Δhel; compare arrow in Figure 15A), GFP+PDspec failed to enrich Δhel-Flag-Dcr-2 in comparison to the input and showed only marginally higher levels than background binding observed for myc-GFP alone (Figure 15A). Similarly, both Loqs-PD and R2D2 recovery were almost completely lost when Δhel-Flag-Dcr-2 was immunoprecipitated (Figure 15C). Taken together, the helicase domain of Dcr-2 is required for interaction with both Loqs-PD and R2D2. Note that co-expression of either R2D2 or Loqs-PD appears to stabilize Dcr-2; mutual *in vivo* stabilization has been reported previously for both the R2D2/Dcr-2 pair (Liu et al., 2003) and for the Loqs-PB/Dcr-1 complex (Forstemann et al., 2005; Liu et al., 2007).

Further co-immunoprecipitation experiments confirmed a clear functional distinction between the Loqs-PB C-terminus, which encompasses the third dsRBD and thus mediates the interaction with Dcr-1 (Forstemann et al., 2005; Ye et al., 2007) and the Loqs-PD-specific C-terminus. Figure 15D shows that the Loqs-PD C-terminus can neither interact with the full-length Flag-Dcr-2, nor with the overexpressed Δhel-Flag-Dcr-2. Interaction of both R2D2 and Loqs-PD with Flag-Dcr-2 but not Δhel-Flag-Dcr-2 was confirmed (Figure 15D).

I tested whether a Δhel-Flag-Dcr-2 protein had a dominant-negative effect on endo-siRNA biogenesis by transfecting the endo-siRNA cell culture reporter with either the full-length form of Flag-Dcr-2 or the Δhel-Flag-Dcr-2 construct. Transfection of truncated Dcr-2 resulted in an approximately twofold increase in reporter fluorescence (Figure 16A). Simultaneous overexpression of Loqs-PD could not counteract this dominant-negative effect, whereas RNAi against R2D2 reduced the effect of Δhel-Flag-Dcr-2 overexpression (Figure 16B). This is consistent with an antagonistic relation between R2D2 and Loqs-PD.

miR-277 is processed by the typical miRNA biogenesis factors Dcr-1 and Loqs, but loaded mainly into Ago2 complexes by the canonical siRNA RLC consisting of Dcr-2 and R2D2 (Förstemann, 2007). Figure 16B and C show the effects of Flag-Dcr2 and Δhel-Flag-Dcr-2 in the miR-277 perfect match reporter cell line. Unlike the endo-siRNA reporter, the miR-277 perfect match reporter does not show a significant difference between overexpression of full-length or truncated Flag-Dcr-2. This is consistent with previously published results, showing that a point mutation in the helicase region of Dcr-2, which prevented dicing but not loading, did not have an effect on the miR-277 perfect match reporter system (Förstemann, 2007).

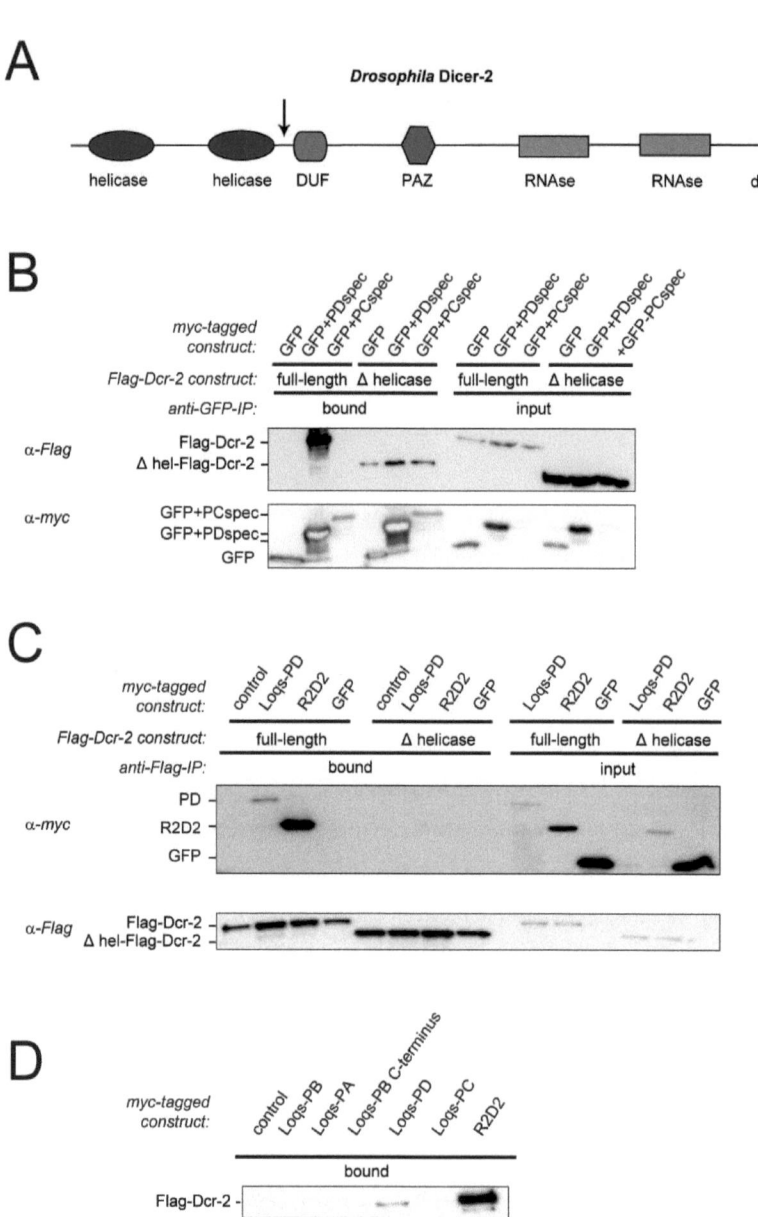

Figure 15: Loqs-PD interacts with the N-terminal helicase domain of Dcr-2 for endo-siRNA silencing (legend continued on p. 82)

(legend Figure 15 continued)
- **A)** Domain architecture of *Drosophila* Dcr-2; blue ovals mark the two DExH/D helicase motifs that are deleted in the Δhel-Flag-Dcr-2 protein; the arrow indicates the position of the forward primer used for cloning the truncated version; DUF = Domain of Unknown Function (probably a double-stranded RNA binding motif); PAZ = "Piwi, Aubergine, Zwille" domain, binds small RNA precursor; RNAse = two RNAseIII type endonycleolytic domains for cleavage of siRNA precursors; dsRBD = double-stranded RNA binding domain
- **B)** Co-immunoprecipitation from *Drosophila* S2 cell extract co-expressing GFP-fusion proteins together with either Flag-Dcr-2 or Δhel-Flag-Dcr2; the GFP proteins in this experiment also contained an N-terminal myc-tag that was used for detection, but GFP-trap beads were used for immunoprecipitation; due to the high amount of immunoprecipitated proteins, some bands in the α-myc Western appear "hollow" because of ECL substrate depletion
- **C)** Co-immunoprecipitation from *Drosophila* S2 cell extract co-expressing myc-tagged Loqs-PD, R2D2 or a pUC18 control together with either Flag-Dcr-2 or Δhel-Flag-Dcr-2; α-Flag agarose was used for IP; myc-GFP served as a control
- **D)** Co-immunoprecipitation from *Drosophila* S2 cell extract co-expressing myc-tagged Loqs isoforms, the Loqs-PB C-terminus, R2D2 or a pUC18 control together with either Flag-Dcr-2 or Δhel-Flag-Dcr-2; α-myc antibody was used for IP; α-Flag antibody was used for Western

(Legend Figure 16; see next page)
(A-B) Effect on 63N1 endo-siRNA cell culture reporter, **(C-D)** effect on miR-277 perfect match reporter
- **A)** Effect of Flag-Dcr-2 (black bars) and Δhel-Flag-Dcr-2 (white bars) overexpression on GFP fluorescence of the endo-siRNA cell culture reporter; Dcr-2 construct expression was combined with Loqs-PD co-expression or an untransfected control
- **B)** Effect of Flag-Dcr-2 (black bars) and Δhel-Flag-Dcr-2 (white bars) overexpression on GFP fluorescence of the endo-siRNA cell culture reporter; Dcr-2 construct expression was combined with RNAi against either mRNAs of Loqs-PD, R2D2 or a control (dsRed)
- **C)** Effect of Flag-Dcr-2 (black bars) and Δhel-Flag-Dcr-2 (white bars) overexpression on GFP fluorescence of the miR-277 perfect match cell culture reporter; Dcr-2 construct expression was combined with Loqs-PD co-expression or an untransfected control
- **D)** Effect of Flag-Dcr-2 (black bars) and Δhel-Flag-Dcr-2 (white bars) overexpression on GFP fluorescence of the miR-277 perfect match cell culture reporter; Dcr-2 construct expression was combined with RNAi against either mRNAs of Loqs-PD, R2D2 or a control (dsRed)

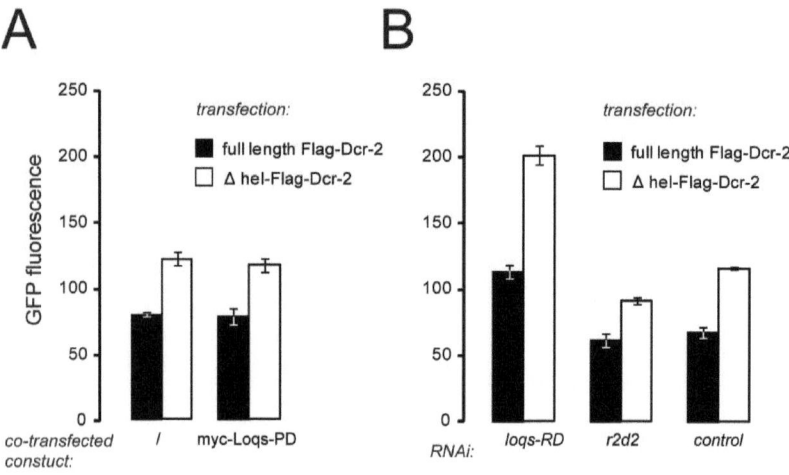
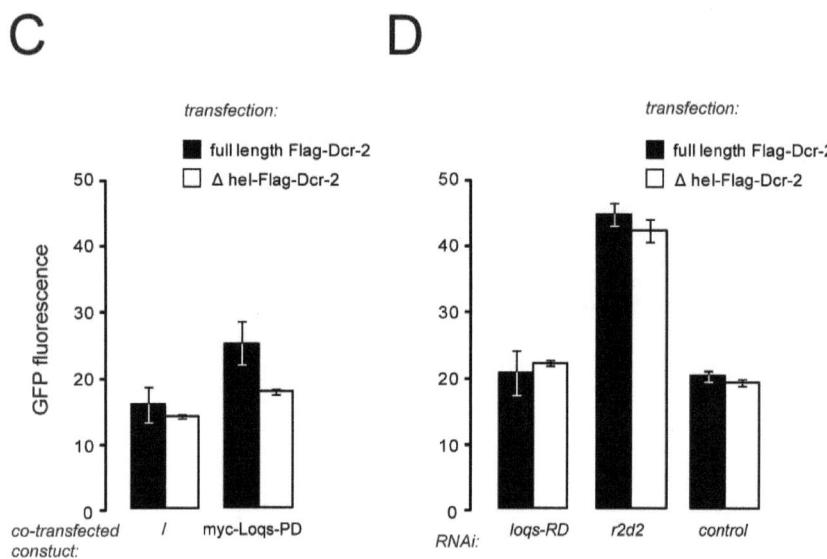

Figure 16: Effect of Δhel-Flag-Dcr-2 overexpression on endo-siRNA and miR-277 perfect match reporter cell lines (legend continued on p. 82)

5.10 R2D2 acts as an antagonist of Loqs-PD in endo-siRNA silencing

A possible explanation for the antagonistic behavior of Loqs and R2D2 is competition for binding to Dcr-2, thereby committing the enzyme to either the endo-siRNA or the exo-siRNA pathway. Reduced co-precipitation of both Loqs-PD and R2D2 with the Δhel-Flag-Dcr-2 construct (Figure 15C) substantiates this hypothesis as both proteins seem to interact with the helicase region of Dcr-2 (Ye et al., 2007; Lim do et al., 2008). Binding of Loqs-PD and R2D2 should be mutually exclusive if the same binding site is used; alternatively, Loqs-PD and R2D2 could simultaneously bind Dcr-2 but induce different conformations.

A recent finding by the Siomi lab (Miyoshi et al., 2010) reports a small amount of R2D2 association together with the Loqs-PD/Dcr-2 complex. I co-expressed a Flag-myc-tagged version of the genomic Loqs-PD construct (compare Figure 6A) together with myc-tagged Loqs isoforms or R2D2 in S2 cells. Upon α-Flag IP I recovered a minor amount of myc-R2D2 associated with Flag-myc-Loqs-PD (Figure 17A, first panel). To reduce the effects of overexpression, I used the PD-specific antibody to precipitate endogenous Loqs-PD. Association of overexpressed R2D2 with endogenous Loqs-PD was barely detectable (Figure 17A, second panel). Co-precipitation of endogenous R2D2 after PD-specific IP was difficult to analyze, since R2D2 and the antibody light chain migrated closely together (Figure 17B). As the amount of R2D2 that co-precipitated with Loqs-PD was sub-stoichiometric even when overexpressed, I favor the hypothesis that binding of Loqs-PD and R2D2 to Dcr-2 is mutually exclusive.

To test whether the competition between Loqs-PD and R2D2 can be observed at the level of endo-siRNA biogenesis, I treated S2 cells with RNAi against combinations of Loqs isoforms together with R2D2. Subsequently, I isolated RNA and probed Northern blots for the long hairpin-derived endo-siRNA *CG4068* B, as well as *bantam* and miR-277. As indicated before (Figure 11A, B), I observed impaired endo-siRNA biogenesis of *CG4068* B after treatment with the *loqs-ORF* RNAi construct against all Loqs isoforms and – even more pronounced – with dsRNA against Loqs-PD specifically (Figure 17C). Simultaneous knock-down of R2D2 led to higher levels of mature endo-siRNA in all combinations. The simplest explanation for this effect is that depletion of R2D2 allows more efficient processing of the long hairpin endo-siRNA precursor by the Loqs-PD/Dcr-2 complex.

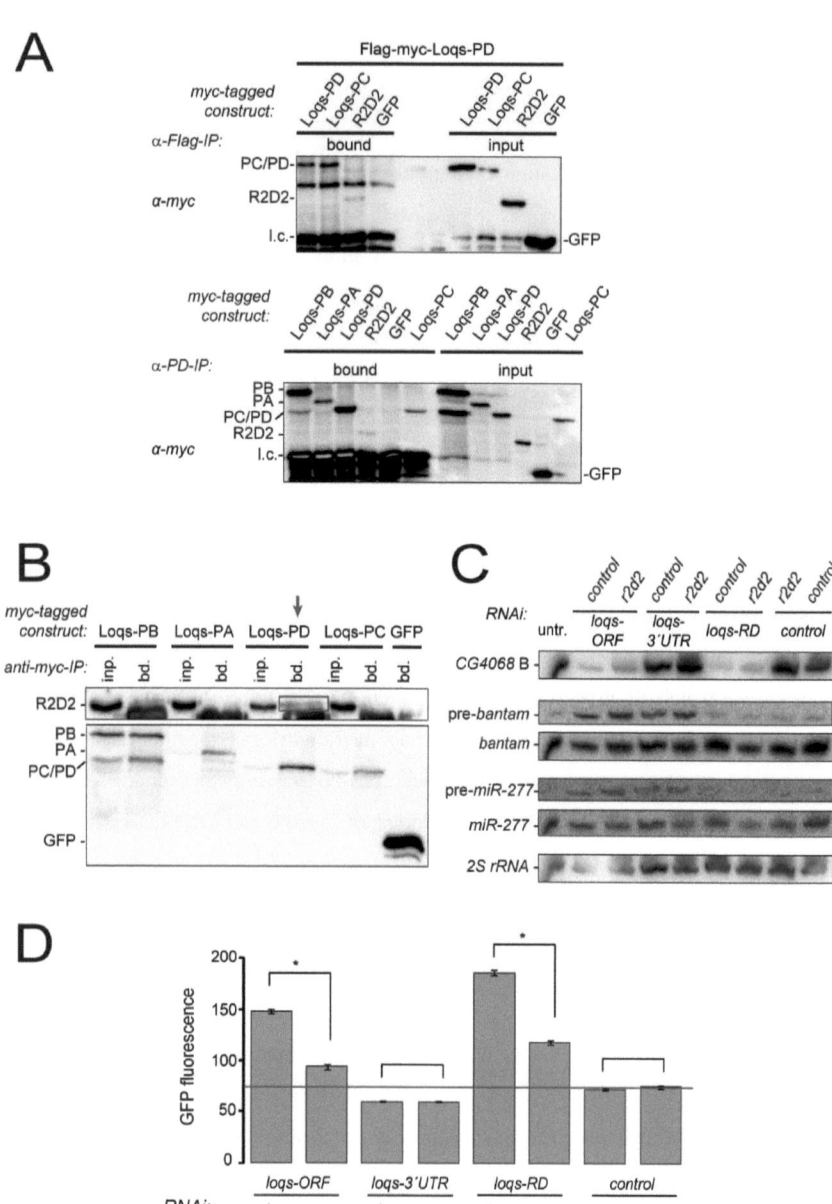

Figure 17: R2D2 minimally associates with Dcr-2 and Loqs-PD and acts as an inhibitor of Loqs-PD in endo-siRNA silencing (legend continued on p. 86)

(legend Figure 17 continued)
A) *Top panel:* Co-immunoprecipitation from *Drosophila* S2 cell extract co-expressing myc-Loqs isoforms or R2D2 together with Flag-myc-Loqs-PC/-PD; α-Flag antibody was used for IP; myc-GFP served as a control; note that Flag-myc-Loqs-PD migrates at the same height as myc-Loqs-PB
Bottom panel: Co-immunoprecipitation from *Drosophila* S2 cell extract prepared after transfection of myc-tagged Loqs isoforms or myc-tagged R2D2; α-Loqs-PD-specific antibody was used to immunoprecipitate endogenous Loqs-PD; myc-GFP served as a control
l.c. = antibody light-chain
B) Co-immunoprecipitation from *Drosophila* S2 cell extract expressing myc-Loqs isoforms; α-myc antibody was used for IP; myc-GFP served as a control; endogenous R2D2 was detected with polyclonal α-R2D2 antibody; red arrow and red square indicate faint band of possible R2D2 co-precipitation above the antibody light-chain band in the Loqs-PD bound fraction
C) Northern blot from *Drosophila* S2 cell extract after RNAi treatment with combinations of Loqs isoforms and R2D2; DsRed dsRNA served as a control for RNAi; a DNA probe against the long hairpin-derived endo-siRNA *CG4068* B and 2´-OMe oligonucleotide probes against *bantam* miRNA and miR-277 were used; 2S rRNA served as a control for loading
D) Effect of RNAi treatment with combinations of Loqs isoforms and R2D2 on GFP expression of the endo-siRNA cell culture reporter; double-stranded RNA directed against DsRed served as a control, RNAi triggers are indicated below the bars; measurement values represent the mean ± SD (n=3) and were normalized to a DsRed/DsRed control (red line); asterisk: p<0.005 (two-tailed t-Test, unequal variance)

I then tested whether enhanced biogenesis of endo-siRNAs led to improved endo-siRNA silencing in the 63N1 cell culture reporter. Figure 17D shows that this is indeed the case, indicating that my observation is true for long hairpin- as well as transgene-derived endo-siRNAs. RNAi against R2D2 had no significant effect on the reporter if Loqs-PD levels were unaltered (RNAi against the 3'UTR or DsRed control). However, simultaneous R2D2 depletion could reduce the impairment of endo-siRNA silencing caused by Loqs-PD depletion (RNAi against all Loqs isoforms or Loqs-PD), indicating that the ratio of Loqs-PD and R2D2 in the cell can influence the efficiency of endo-siRNA silencing.

5.11 The role of Loqs-PD in exo-siRNA silencing

My experiments indicate that R2D2 is not required for endo-siRNA dependent silencing in S2 cells but rather acts as a competitor of Loqs-PD. This is consistent with previous cell culture studies (Czech et al., 2008; Ghildiyal et al., 2008; Kawamura et al., 2008; Okamura et al., 2008a; Okamura et al., 2008b; Zhou et al., 2009) but inconsistent with results from a recent study conducted in mutant fly tissue (Marques et al., 2010). They propose, that Loqs-PD is necessary for processing of both exo-siRNAs and endo-siRNAs and that a common Dicer-2/R2D2 RLC exists to funnel precursors into Ago2 RISCs. Given the obvious discrepancy for the R2D2 role in endo-siRNA silencing between flies and S2 cells, I looked for evidence of

Loqs-PD involvement in exo-siRNA biogenesis in cell culture. S2 cells were treated with RNAi triggers against small RNA silencing components. To visualize exo-siRNA silencing efficiency cells were then soaked with dsRNA against GFP or a control and transfected with the pKF63 expression construct for myc-GFP (Figure 18A). Priming of Ago2 RISCs with dsGFP will decrease in efficiency, if the first knock-down depletes an essential component for exo-siRNA silencing. This will result in higher GFP levels after pKF63 transfection. GFP levels were efficiently repressed to 3% in dsGFP treated cells compared to the control, if non-essential components were targeted in the first step (Figure 18B; compare *luciferase* RNAi). Depletion of the canonical exo-siRNA pathway components Dcr-2 and Ago2 reduced the efficiency of exo-siRNA silencing, resulting in a GFP fluorescence of 25% or 35% compared to the control. This indicates that the assay can be used for exo-siRNA silencing analysis. Of the two dsRBDs, *r2d2* RNAi impaired exo-siRNA silencing (10% compared to control), although the effect was not as pronounced as for *dcr-2* or *ago2*. Loqs-PD depletion, on the other hand, slightly but significantly enhanced silencing of GFP ($p < 0.005$). Taken together the reporter suggests that R2D2 enhances exo-siRNA silencing while Loqs-PD is not required and may even act as an inhibitor of R2D2 function.

Previously, a non-canonical RNA-dependent RNA-polymerase (RdRP) was observed by biochemical evidence in *Drosophila* and recently identified as D-elp-1, a subunit of the PolII elongator complex (Lipardi et al., 2001; Lipardi et al., 2009). RdRP activity in other organisms is known to produce secondary siRNAs (Simmer et al., 2002; Pak et al., 2007; Sijen et al., 2007). However, no other study has so far observed any evidence for secondary siRNAs or spreading of siRNA responses in *Drosophila melanogaster*. Therefore I included RNAi against D-elp-1 in the experiment but could not observe a significant decrease of siRNA silencing efficiency after depletion of D-elp-1. Thus, I could not confirm any RdRP-like activity of D-elp-1 in my reporter system. However, efficiency of RNAi against D-elp-1 cannot be tested in Western blotting, since there is no commercially available antibody for D-elp-1 detection. Efficiency of RNAi against D-elp-1 might be assessed on the mRNA level by quantitative PCR.

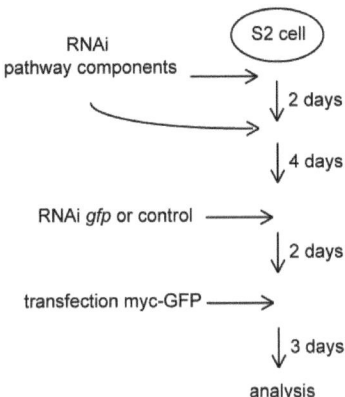
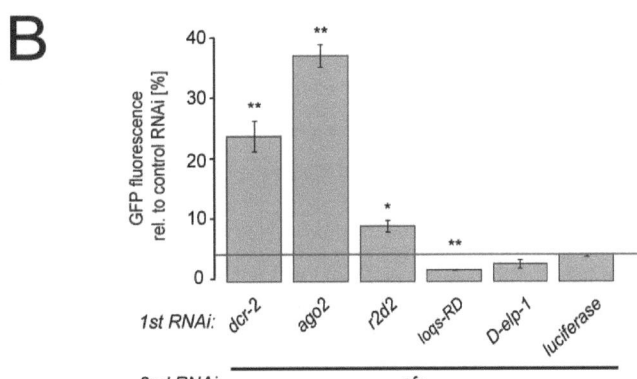

Figure 18: Loqs-PD is not necessary for exo-siRNA dependent silencing
A) Flow diagram for exo-siRNA silencing assay; S2 cells were treated twice by RNAi against small RNA silencing factors; Ago2 RISCs were then primed with RNAi triggers against *gfp* or a *luciferase* control and transfected with an expression plasmid for GFP (pKF63); GFP levels were measured by flow cytometry
B) Effect of R2D2 and Loqs-PD on the exo-siRNA silencing assay; RNAi triggers are indicated below the bars; GFP fluorescence relative to a control treated with dsRNA against luciferase in the 2nd step of RNAi; the red bar marks GFP levels if the exo-siRNA silencing system is unaltered; ** $p<0.005$, * $p<0.02$ (two-tailed t-Test, unequal variance)

5.12 Multimerization and competition of Loqs isoforms for Dcr-2 binding

As shown in Figure 14B, overexpression of other Loqs isoforms leads to impaired endo-siRNA silencing. By transfecting endo-siRNA reporter cells with increasing amounts of Loqs isoforms or only the Loqs-PB C-terminus (encompassing the third dsRBD of Loqs-PB), I tested for a correlation between the amount of transfected expression plasmid and an increase in GFP-expression (

Figure 19A). The expression of the Loqs-PB C-terminus, which is responsible for Dcr-1 interaction, did not have a significant influence on the reporter, nor did Loqs-PD. In contrast, I saw a strong correlation between the quantity of transfected Loqs-PA or -PB with impaired endo-siRNA silencing. The effect of Loqs-PC was considerably weaker; this may however be due to lower expression levels of the construct (see Appendix 4). The de-repression of the reporter was not influenced by the epitope tag, since transfection with Flag-myc-tagged constructs caused a comparable effect in the endo-siRNA cell culture reporter (

Figure 19B). Moderate overexpression levels of Loqs-PD here even lead to hyperrepression of GFP levels. Could the dominant-negative effect be explained by multimerization of overexpressed Loqs isoforms together with endogenous Loqs-PD and the formation of dsRBP complexes unsuitable for silencing? I overexpressed Flag-myc-Loqs-PB together with myc-tagged Loqs isoforms or R2D2 and analyzed the bound fraction after α-Flag-IP (

Figure 19C, left panel). While recovery of Loqs-PA/-PB/-PC was high, Loqs-PD associated only to a moderate extent with Loqs-PB, and R2D2 was not detected in the bound fraction at all. The result indicates that overexpressed Loqs isoforms can oligomerize, but that R2D2 does not form a complex with Loqs-PB. Analogous co-immunoprecipitation from cells expressing the Flag-myc-Loqs-PD$_{genomic}$ construct yielded a barely detectable amount of R2D2 in the bound fraction, as well as considerable Loqs-PA and -PC recovery (

Figure 19C, right panel; Loqs-PB association was not quantifiable, since myc-Loqs-PB and Flag-myc-Loqs-PD migrate together).

To further analyze the association of myc-tagged Loqs variants with Loqs-PD I used the Loqs-PD-specific antibody to precipitate endogenous Loqs-PD (see Figure 17A, bottom panel). The result confirmed the finding that an interaction of Loqs-PD with all other isoforms as well as

R2D2 in *Drosophila* S2 cells is at least possible. However, these interactions may have been forced by overexpression and might be indirect. I therefore tested for oligomerization of Loqs-PD and other Loqs-isoforms at endogenous levels by immunoprecipitation with our Loqs-PD specific antibody, followed by Western blotting with α-Loqs monoclonal antibody. This did not reveal any significant recovery of Loqs-PB or Loqs-PA associated with Loqs-PD (

Figure 19D, α-Dcr-1 IP was used as a control to mark Loqs-PB precipitation). Taken together, my experiments indicate that endogenous Loqs-PD does not form complexes with other Loqs isoforms to a significant extent but can be induced to do so upon overexpression.

If overexpression of other isoforms impairs endo-siRNA silencing on the level of Loqs-PD/Dcr-2 association, then endo-siRNA biogenesis should be disturbed. Loqs-PD overexpression from both cDNA- and genomic DNA-derived expression constructs did not change the levels of the long hairpin-derived endo-siRNA *CG4068* B. Expression of full-length myc-Loqs-PB/-PA or Loqs-PC (either full-length or reconstituted L1L2+PCspec) led to a reduction of mature *CG4068* B (

Figure 19E). In accordance with my reporter cell experiments (Figure 14B, C) truncated Loqs, lacking only the PD-specific amino acid sequence (L1L2), caused the same impairment of long hairpin endo-siRNA biogenesis as the Loqs-PC isoform (

Figure 19E). In summary, multimerization induced by overexpression of individual Loqs isoforms perturbs the balance of the other, endogenous dsRBD-proteins and impairs the function of the small RNA silencing system.

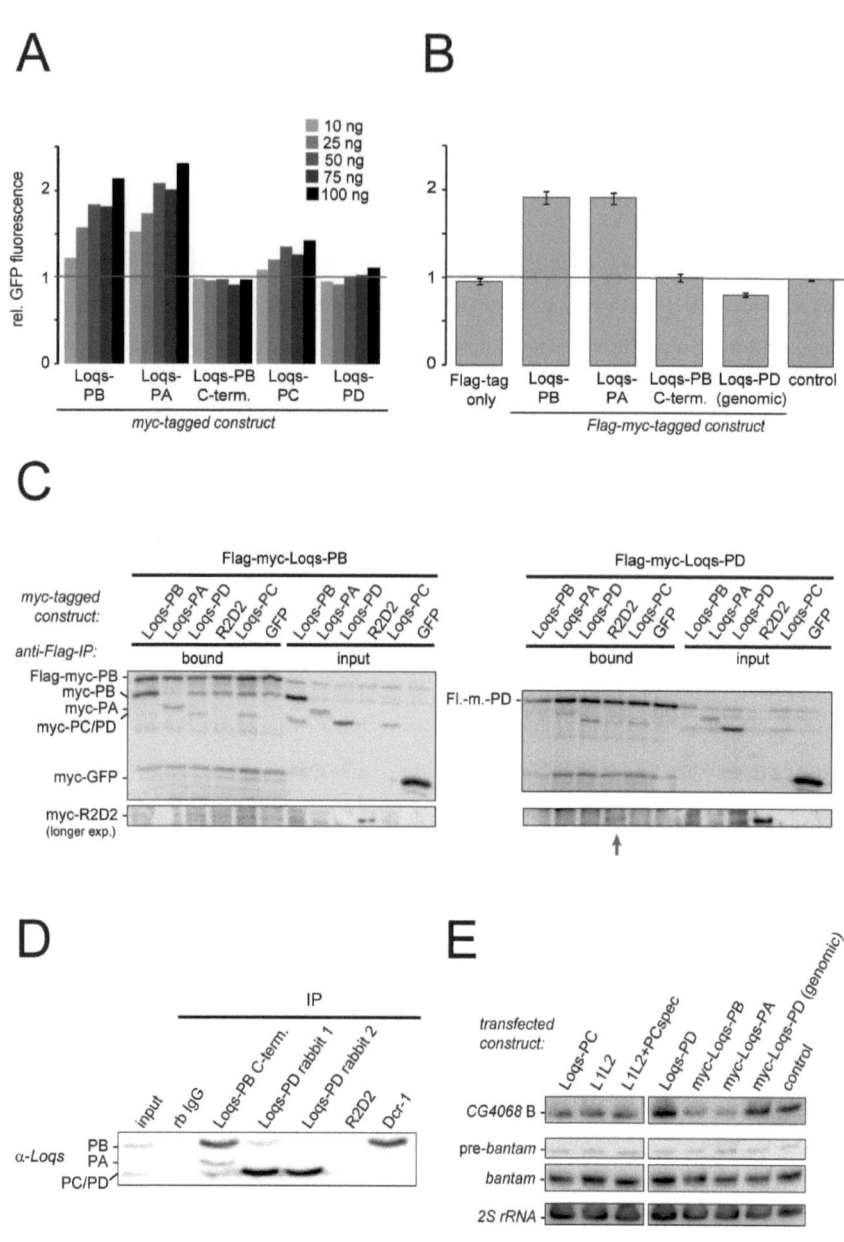

Figure 19: Multimerization and competition of Loqs isoforms for Dcr-2 binding (legend continued on p. 92)

(legend Figure 19 continued)
A) Correlation of GFP fluorescence increase in the endo-siRNA cell culture reporter with the amount of transfected plasmid; the transfected myc-tagged construct is indicated below the bars; cells were transfected with 10 ng, 25 ng, 50 ng, 75 ng and 100 ng of the respective expression plasmid; the values were normalized to a pUC18 control (red line); Loqs-PB C-term. = expression construct for the 3rd dsRBD of Loqs-PB
B) Effect of Flag-myc-tagged Loqs-isoforms on the endo-siRNA cell culture reporter; transfected plasmids are indicated below the bars; mock-transfection (red horizontal line) and an expression construct for the Flag-tag only served as controls; measurement values represent the mean ± SD (n=3)
C) *Left panel:* α-myc Western blot after immunoprecipitation of Flag-myc-Loqs-PB from *Drosophila* S2 cell extracts co-expressing myc-Loqs isoforms or myc-R2D2; α-Flag antibody was used for IP; myc-GFP served as a control;
Right panel: α-myc Western blot after immunoprecipitation of Flag-myc-Loqs-PD from *Drosophila* S2 cell extracts co-expressing myc-Loqs isoforms or myc-R2D2; α-Flag antibody was used for IP; myc-GFP served as a control; note that Flag-myc-Loqs-PD co-migrates with myc-Loqs-PB during SDS-PAGE;
bottom panels show a longer exposure to detect faint R2D2 bands; red arrow indicates co-precipitated R2D2
D) Detection of endogenous Loqs protein from S2-cell extract after immunoprecipitation with α-Dcr-1, α-Loqs-PB C-terminus and α-Loqs-PD; rb IgG and α-R2D2 served as controls; α-Loqs monoclonal antibody was used for detection of endogenous Loqs isoforms
E) Northern blot from *Drosophila* S2 cell extract overexpressing Loqs isoforms; transfection with pUC18 served as a control; L1L2 = Loqs truncation lacking the PD-specific amino acid sequence, L1L2+PCspec = reconstituted Loqs-PC; myc-loqs-PD (genomic) = expression construct derived from genomic DNA; DNA probes against the long hairpin-derived endo-siRNA *CG4068* B (Okamura et al., 2008c) and against *bantam* miRNA were used; 2S rRNA served as a control for loading

5.13 Transcriptional vs. post-transcriptional gene silencing

Endogenous RNA targets of the endo-siRNA pathway become more abundant upon depletion of Dcr-2 or Ago2. The simplest interpretation of these results is that endo-siRNAs, like exo-siRNAs, induce the post-transcriptional degradation of mRNAs, and this capacity has indeed been demonstrated (Chung et al., 2008; Czech et al., 2008; Ghildiyal et al., 2008; Kawamura et al., 2008; Okamura et al., 2008b). However, a transcriptional component of silencing may be present as well. Transcriptional silencing by chromatin remodeling has previously been reported in the yeast *S. pombe*, mammals and plants (Girard et al., 2008). I treated endo-siRNA reporter cells with dsRNA against heterochromatin protein 1 (HP-1) and probed for efficient depletion by Western blotting with a HP-1 specific antibody (Figure 20B). Since HP-1 is chromatin-associated I used extraction buffers with low and high $MgCl_2$ conditions but could not detect a significant difference in the extraction efficiency. Even after efficient RNAi, however, I could observe no effect on the 63N1 reporter cell line (Figure 20A).

p68/Lip is the *Drosophila* homolog of the mammalian RNA helicase P68 and was described to function in RNA export and rapid removal of transcripts from sites that need chromatin remodeling for transcriptional silencing (Buszczak et al., 2006). p68/Lip can be found in a complex with Ago2 and is required for efficient RNAi (Ishizuka et al., 2002), hinting at a possible transcriptional element of silencing. It can be found in a complex with dFMR1, the Drosophila homolog of the human Fragile X Mental Retardation Protein 1 (Ishizuka et al., 2002). RNAi against FMR1 led to a slight repression of GFP levels (Figure 20A), p68/Lip to a slight increase (Figure 20C; 63N1 cell line). Taken together, I could detect no significant dependence of endo-siRNA silencing on transcriptional components.

5.14 Comparison between two clonal cell lines: 63-6 and 63N1

The 63N1 cell line has mounted a *bona fide* endo-siRNA response against the integrated transgene, but the extent to which this occurs varies between clones: As indicated before, our laboratory cell culture stock comprises several GFP expressing cell lines derived independently from the same stock of parental S2 cells and transfected with the same expression plasmid. In a previous publication, the cell line (called 63-6) showed only marginal response to depletion of Dcr-2, Loqs and Ago2 (Förstemann, 2007). I re-examined this cell line and could corroborate that there is a minor, but significant increase in GFP-levels upon depletion of Dcr-2 (1.2-fold, ± 0.03, $p<0.01$; Figure 20C). The level of GFP fluorescence is generally higher in the 63N1 cells. This suggests that one potential difference between the two cell lines is the number of plasmid copies that have integrated in the genome. However, the 63-6 cell line reacted with a strong increase of GFP levels to p68/Lip RNAi while the 63N1 line only shows a slight reaction (Figure 20C).

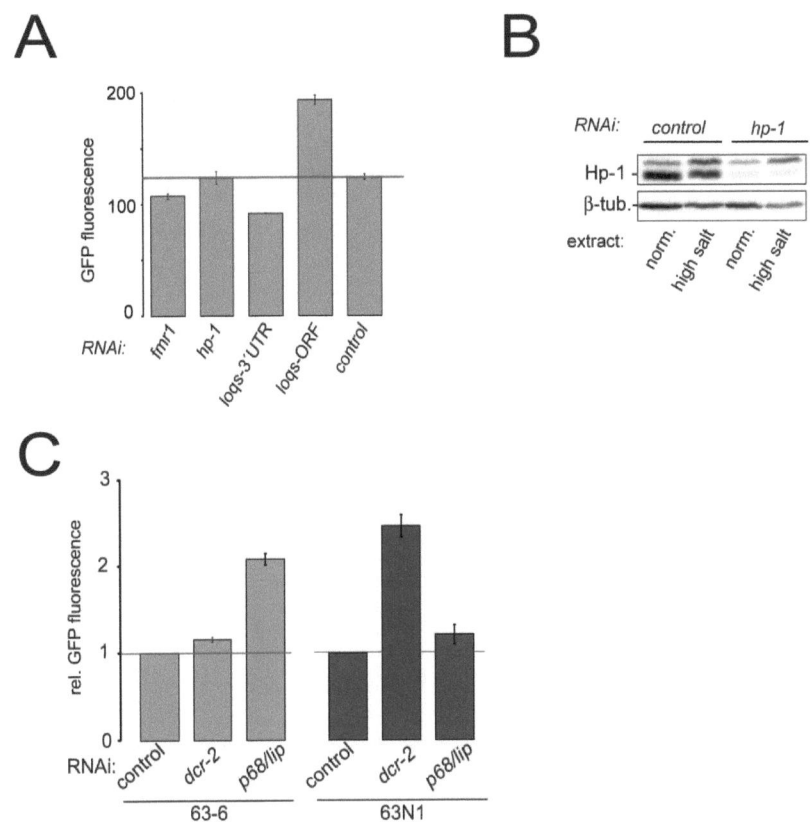

Figure 20: Endo-siRNAs: Transcriptional *versus* post-transcriptional silencing?
A) Effect of RNAi against factors associated with transcriptional gene silencing in the endo-siRNA cell culture reporter; horizontal red line indicates no change compared to a dsRed RNAi control; measurement values represent the mean ± SD (n=3)
B) α-HP-1 Western blot to verify efficient depletion of HP-1; α-tubulin Western served as a loading control; extraction buffers with normal and high $MgCl_2$ conditions (see Materials 4.1.15) were used
C) Comparison of RNAi effects in two GFP-expressing cell lines, 63-6 and 63N1; they represent two independently derived clones from the same parental cells using the same expression plasmid; RNAi triggers are indicated below the bars; measurement values represent the mean ± SD (n=3) normalized against a dsRed RNAi control

5.15 Loqs-PD associates with Dcr-2 *in vivo*

To test whether the role of Loqs-PD for endo-siRNA silencing could be substantiated *in vivo* I created transgenic flies expressing myc-tagged isoforms under the control of the UAS/Gal4-system. Therefore I sub-cloned myc-tagged Loqs isoforms into pUAST-plasmids and tested the constructs in the 63N1 endo-siRNA cell culture reporter. Only upon co-transfection of a tubulin-Gal4 driver did the reporter respond with de-repression of GFP fluorescence to both pUAST-PC and -PD constructs (

Figure 21, left panel). Conditional GFP-expression in S2 cells demonstrated functionality of the tubulin-Gal4 driver plasmid (

Figure 21A, right panel). I confirmed expression in cell culture by Western blotting: Loqs-PC showed only moderate expression while Loqs-PD was highly expressed (

Figure 21B). The high expression level of Loqs-PD may account for the observed de-repression of the 63N1 reporter and is consistent with the previously indicated correlation of transfected plasmid and reporter fluorescence (

Figure 19A). Expression of UAST-Loqs-PA in Western blotting (

Figure 21C) and the 63N1 cell culture reporter (data not shown) was similarly checked. Loqs-PC and -PD plasmids were sent for embryo injection and several homozygous fly lines were recovered (see Results 5.17, Table 4). Together with an analogous construct for Loqs-PB (Park et al., 2007), the flies were crossed to a tubulin-Gal4 driver-line. Western blotting again detected expression of Loqs-PB, -PC and -PD in the resulting offspring flies (

Figure 21D), with Loqs-PD being most prominently expressed. Co-immunoprecipitation experiments with protein extracts from these offspring flies confirmed that myc-tagged Loqs-PD – but not Loqs-PB or -PC – interacts with Dcr-2 (

Figure 21E).

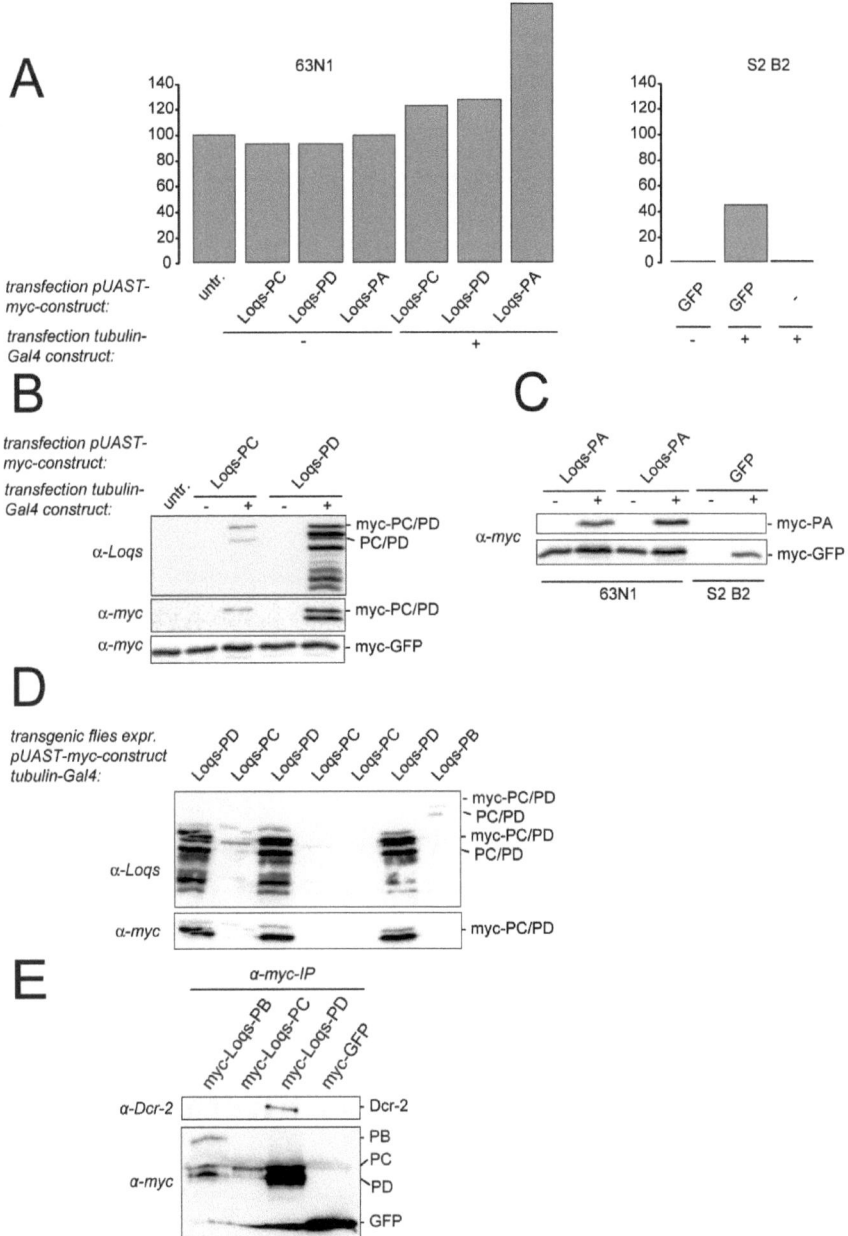

Figure 21: Expression of Loqs-isoforms under the control of the UAS/Gal4 system (legend continued on p. 97)

(legend Figure 21 continued)
A) *Left panel:* Effect of UAST-myc-Loqs-PA/-PC/-PD expression constructs co-transfected with (+) or without (-) a tubulin-Gal4 driver plasmid into 63N1 cell culture reporter cells
Right panel: Test of the driver-plasmid by conditional expression of a UAS-myc-GFP plasmid in S2 cells
Results of a single experiment are shown
B) Western blot to confirm expression of UAST-myc-Loqs-PC and -PD in 63N1 cell culture cells (+); samples without co-expression of Gal4 served as a control (-); α-loqs Western shows degradation products of highly expressed UAS-myc-Loqs-PD; myc-GFP expression of the 63N1 cell line was not significantly altered by transgene expression
C) Western blot to confirm expression of UAST-myc-Loqs-PA in 63N1 cell culture cells (+); samples without co-expression of Gal4 served as a control (-); two different UAST-myc-Loqs-PA clones were transfected; myc-GFP expression of the 63N1 cell line was not significantly altered by transgene expression; induced UAS-myc-GFP expression in S2 cells confirmed functionality of the system
D) Western blot for expression of UAST-myc-Loqs-PC and -PD in transgenic flies; several homozygous transgenic lines were crossed with a tubulin-Gal4 driver line and offspring were assayed for transgene expression; a previously published UASP-myc-Loqs-PB expression construct (Park et al., 2007) served as a control; α-loqs Western shows degradation products of highly expressed UAST-myc-Loqs-PD
E) Co-immunoprecipitation of endogenous Dcr-2 with myc-tagged Loqs isoforms from transgenic *Drosophila* fly extract; the blot was re-probed with α-myc antibodies to verify successful immunoprecipitation of the Loqs proteins; the smaller size bands in the myc-Loqs-PB lane likely represent degradation products; myc-GFP expressing flies were used as a non-specific control

5.16 Loqs-PD is essential for endo-siRNA biogenesis *in vivo*

The next step was to analyze the role of Loqs-PD for living animals. A transposon insertion mutant of Loqs (loqs^{f0079}; Forstemann et al., 2005) shows reduced expression of all Loqs protein isoforms and can be used to approximate the effect of a complete Loqs knock-out allele. As could be anticipated, both endo-siRNA and miRNA biogeneses were disrupted in loqs^{f0079} flies (

Figure 22B). It has been described previously that Loqs-PB expression in a Loqs^{KO} background was sufficient to rescue miRNA biogenesis (Park et al., 2007). However, it did not suffice to reconstitute processing of the long hairpin-derived endo-siRNA *CG4068* B (

Figure 22B). In contrast, flies expressing additional Loqs-PD had even higher levels of mature endo-siRNA *CG4068* B than the wildtype (OrR), while *bantam* biogenesis was unaffected (

Figure 22B). Loqs-PC expression did not significantly affect biogenesis of the analyzed small RNAs (data not shown).

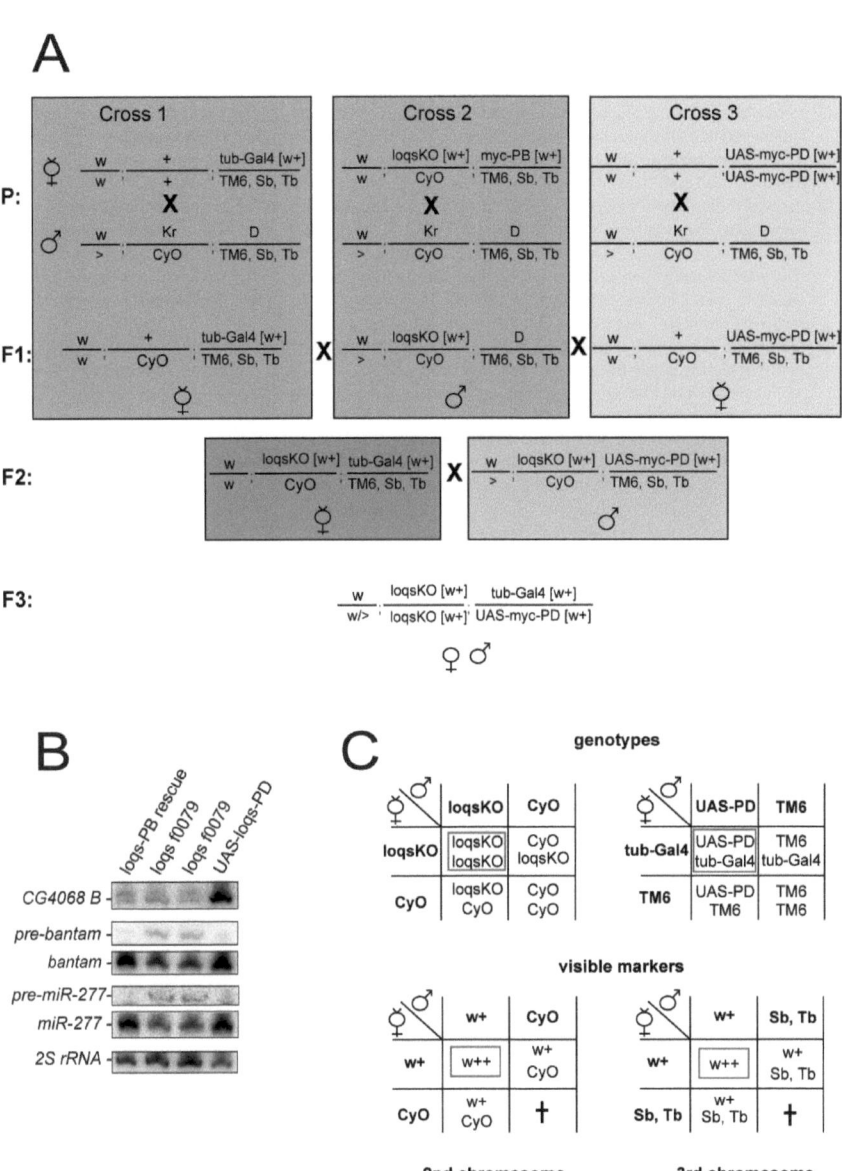

Figure 22: Transgenic flies expressing only the Loqs-PD isoform (legend continued on p. 99)

(legend Figure 22 continued)
A) Mating scheme for recovery of Loqs-PD rescue flies with abbreviated phenotypes (see Materials and Methods 4.1.10 and 4.2.5.2.3); virgins of a tubulin-Gal4 driver line (cross 1; marked in red), the Loqs-PB rescue line (cross 2; marked in blue) and homozygous UAST-Loqs-PD transgenic flies (cross 3; marked in yellow) were mated with male double-balancer flies; offspring from cross 1 and 3 were both mated with F1 males from cross 2, yielding F2 flies (marked in violet and green); the third generation should produce homozygous Loqs^{KO} flies with an active tub-Gal4/UAST-Loqs-PD expression system
B) Northern blot from *Drosophila* transgenic fly extract; loqs-PB rescue flies (Park et al., 2007) and Loqs-PD expressing offspring from a tubulin-Gal4/UAST-Loqs-PD cross were used; loqs^{f0079} flies (Forstemann et al., 2005) have reduced (5-fold in somatic tissue and up to 40-fold in ovaries) Loqs expression levels due to a BiggyBac transposon insertion 57 nucleotides upstream of the *loqs* transcription start site; a DNA probe against the long hairpin-derived endo-siRNA *CG4068* B and 2´-OMe oligonucleotide probes against *bantam* miRNA and miR-277 were used; 2S rRNA served as a control for loading
C) Schematic overview over possible offspring in the F3 generation (see A); only relevant chromosomes 2 and 3 are depicted; upper panel shows possible combinations of phenotypes, lower panel the corresponding visible phenotypic markers in the adult fly; crosses mark non-viable individuals with homozygous balancer chromosomes; red rectangles mark intended features of F3 offspring

w^+ = gene for red eye color (intensity is additive); CyO = "Curly of Oyster", curly wings; TM6, Sb, Tb = TM6 balancer chromosome with Sb (stubble = short thoracic bristles) and Tb (tubby = segmentation phenotype with short larval form) as phenotypic markers; "<" represents male Y-chromosome

In a subsequent step I made crosses to obtain a Loqs-PD rescue strain (

Figure 22A), that is flies expressing only Loqs-PD in a Loqs^{KO} background analogous to the existing Loqs-PB rescue strain (Park et al., 2007). According to Mendelian laws there is a 1/16 probability for correct F3 offspring (

Figure 22C). Since balancer chromosomes carry recessive lethal mutations, individuals with homozygous balancer chromosomes are non-viable. Therefore 1/9 of surviving offspring should have the intended genotype (

Figure 22C). Table 2 indicates, however, that no PD-rescue flies carrying a homozygous Loqs^{KO} could be recovered. It has been reported (Park et al., 2007) that only a small percentage of homozygous Loqs^{KO} flies survive past pupal stage, due to a developmental defect caused by lack of miRNAs. Adult flies are unhealthy and die shortly after eclosion (Park et al., 2007). Loqs-PB expression in the Loqs-PB rescue flies reconstitutes the miRNA system and normalizes survival rate. No viable Loqs-PD rescue flies suggest that Loqs-PD cannot substitute for Loqs-PB in miRNA biogenesis.

Table 2: Cross to obtain flies expressing Loqs-PD in a Loqs^{KO} background; F3 offspring
Red rectangle marks correct phenotype of Loqs-PD rescue flies; note that no homozygous Loqs^{KO} flies could be recovered, even though there is a 1/3 probability; ratio of heterozygous Loqs^{KO} flies essentially as expected

phenotype offspring	non-CyO non-sb	non-CyO sb	CyO non-sb	CyO sb

# offspring	0	0	156	252
experimental ratio	/	/	1,00	1,62
expected ratio	1	2	2	4

5.17 Future perspective

The next step will be to design a cross that will yield viable offspring of Loqs-PD rescue flies, potentially in a tissue-specific expression system and to extend the experiment to Loqs-PA and Loqs-PC rescue flies. Table 3 lists the UAS/Gal4 expression plasmids, Table 4 shows the potential homozygous fly stocks now available.

Table 3: pUAST expression plasmids for myc-tagged Loqs-PA/-PC/-PD

Plasmid	abbreviation	stock number
pUAST+myc-Loqs-PC	pEH 47	113
pUAST+myc-Loqs-PD	pEH 48	114
pUAST+myc-Loqs-PA	pEH 51	126

Table 4: Overview over UAS-myc-Loqs-PC and -PD transgenic flies
Stock number 318 was used in cross to obtain Loqs-PD rescue flies (see Figure 22A); eye phenotype marks intensity of transgene expression, no remark indicates eye color similar to wildtye flies; mapped and verified stocks are indicated

Original name	Eye phenotype	New Name	Mapped location	Stock number
E PC M 17;38+37	dark red	UAS-Loqs-PC 1	3L; 5241560	300
K PC M 73		UAS-Loqs-PC 2		302
PC 11;3+11		UAS-Loqs-PC 3		303
PC V 12;4+12		UAS-Loqs-PC 4		304
PC V 73;32+36	m red; f dark orange	UAS-Loqs-PC 5		305
PC 50;17+23	m red; f dark orange	UAS-Loqs-PC 6		306
PC 63;35+31	m dark red; f red	UAS-Loqs-PC 7		307
PC V 3;50		UAS-Loqs-PC 8		308
PC M 28;45	light red	UAS-Loqs-PC 9		309
PC M 55; 46+39	varying eye color (red-orange)	UAS-Loqs-PC 10		310
PC V 4; 1+10		UAS-Loqs-PC 11		311
PC V 51; 17+38		UAS-Loqs-PC 12		312

M PC		UAS-Loqs-PC 13	2R; 8519550	330

(Table 4 continued)

Original name	Eye phenotype	New Name	Mapped location	Stock number
B PD M 68		UAS-Loqs-PD 1	3R; 217147	313
A PD V 68		UAS-Loqs-PD 2		314
C PD V 68	dark red	UAS-Loqs-PD 3		315
D PD M 5		UAS-Loqs-PD 4	2L; 5999610	316
G PD V 30		UAS-Loqs-PD 5		317
H PD 53	light red	UAS-Loqs-PD 6	3L; 20488501	318
I PD M 68		UAS-Loqs-PD 7		319
PD V 27;12+3		UAS-Loqs-PD 8		320
PD V 30;4+13		UAS-Loqs-PD 9		321
PD V 35;5+14	dark red	UAS-Loqs-PD 10		322
PD V 38;15+17		UAS-Loqs-PD 11		323
PD V 54;23		UAS-Loqs-PD 12		324
PD V 69;26	dark red	UAS-Loqs-PD 13		325
PD V 78;27	light red	UAS-Loqs-PD 14		326
PD V 49;16+8	orange	UAS-Loqs-PD 15		327
PD M 5; 40+37	dark red	UAS-Loqs-PD 16		328
PD V 68; 25+31		UAS-Loqs-PD 17		329

To further study the complex interplay of small RNA silencing pathways in cell culture, I designed domain-swapped constructs listed in Table 5. Stable cell lines derived from these constructs are shown in Appendix 9. Together they will be a helpful tool in the functional analysis of the dsRBDs of Loqs and R2D2.

GFP-fusion constructs for Loqs-PC- and Loqs-PD-specific C-termini are listed in Table 6, Dcr-2 expression constructs in Table 7.

Table 5: Overview over domain-swapped Loqs/R2D2 constructs

dsRBD elements of Loqs are shaded in orange, dsRBD elements of R2D2 in green; Loqs-PDspec C-terminus is marked in red, Loqs-PCspec C-terminus in blue; "isoform dsRBD 2" indicates, if 46 aa sequence of Loqs-PB was included or not (PA); PCR-derived point mutations as well as subsequent amino acid exchange are indicated for a subset of constructs [nucleotide position refers to *loqs* or *r2d2* genomic regions (see http://flybase.org/reports/FBgn0032515.html) for comparability]; ligation-dependent addition of two amino acids (glycine and lysine) is indicated by a "+"

epitope-tag	dsRBD 1	Ligation	dsRBD 2	Ligation	dsRBD 3/ C-term.	isoform dsRBD 2	comments	abbreviation	Stock #
	R1		R2	+	L3			pEH 1	64
	L1	+	R2		RC-term			pEH 2	65
	L1		L2	+	RC-term	PB		pEH 4	68
	R1	+	L2		L3	PB	nt 892 A→G (aa T→A) nt 1374 T→C (aa L→P)	pEH 5	69
myc	R1		R2		RC-term			pEH 7	71
myc	L1		L2		L3	PA		pEH 8	73
myc	R1		R2	+	L3		nt 517 A→G (aa E→G)	pEH 10	75
	R1	+	L2	+	RC-term	PA		pEH 12	76
	L1		L2		L3	PB		pEH 16	80
myc	L1		L2		L3	PB	nt 1031 T→C (aa F→L)	pEH17	81
	R2		R3		RC-term			pEH 18	82
myc	L1		L2		/	PA	nt 813 C→T (aa P→L) nt 1178 G→A (aa G→S) nt 1191 G→A (aa G→E)	pEH 19	84
	L1		L2		/	PA	nt 813 C→T (aa P→L) nt 1178 G→A (aa G→S) nt 1191 G→A (aa G→E)	pEH 22	87
myc	L1		L2		L3	PA	nt 851 G→C (aa G→R)	pEH 24	89
	L1		L2		L3	PA	nt 851 G→C (aa G→R)	pEH 25	90
myc	L1		L2	+	RC-term	PA	nt 851 G→C (aa G→R) nt 1172 A→G (aa I→V)	pEH 30	95
myc	L1		L2	+	RC-term	PB	nt 937 T→C (aa I→T)	pEH 31	96
	L1		L2		/	PA	truncated Loqs	pEH 33	99
	L1		L2	+	PCspec	PA	reconstituted PC	pEH 34	100
	L1		L2		PCspec		PC isoform	pEH 35	101
myc	L1		L2		PDspec		PD isoform	pEH 37	103
myc	L1		L2		PCspec		PC isoform	pEH 38	104
	L1		L2		PDspec		PD isoform	pEH 39	105
myc					PDspec		PDspec only	pEH 40	106
	R1		R2	+	PCspec			pEH 41	107
	R1		R2		/		truncated R2D2	pEH 42	108
	L1		L2	+	PDspec	PA	reconstituted PD	pEH 44	110
	R1		R2	+	PDspec			pEH 45	111

Table 6: Overview over GFP-fusion proteins
Loqs-PDspec C-terminus is marked in red, Loqs-PCspec C-terminus in blue; ligation-dependent addition of two amino acids (glycine and lysine) is indicated by a "+"

epitope-tag	fusion protein	Ligation	C-terminus	abbreviation	Stock #
	GFP	+	PCspec	pEH 43	109
	GFP	+	PDspec	pEH 46	112
myc	GFP	+	PDspec	pEH 49	115
myc	GFP	+	PCspec	pEH 50	116

Table 7: Overview over Flag-Dcr-2 expression proteins
For Dcr-2 ΔdsRBD see Appendix 8

epitope-tag	fusion protein	abbreviation	Stock #
Flag	Dcr-2 full-length	pEH 53	128
Flag	Dcr-2 Δhel	pEH 54	129
Flag	Dcr-2 ΔdsRBD	pEH 55	130

However, a definite answer about the affinity of individual dsRBDs can only be obtained by *in vitro* studies with recombinantly expressed and purified proteins. I started to optimize conditions for expression and affinity purification of GST- or His$_6$-tagged Loqs and R2D2. Initial promising expression and GST-agarose purification of GST-Loqs-PA, recombinantly expressed in the BL21 *E. coli* strain, is shown in Figure 23A and B, respectively. This project is now continued and expanded in the lab by Stephanie Fesser.

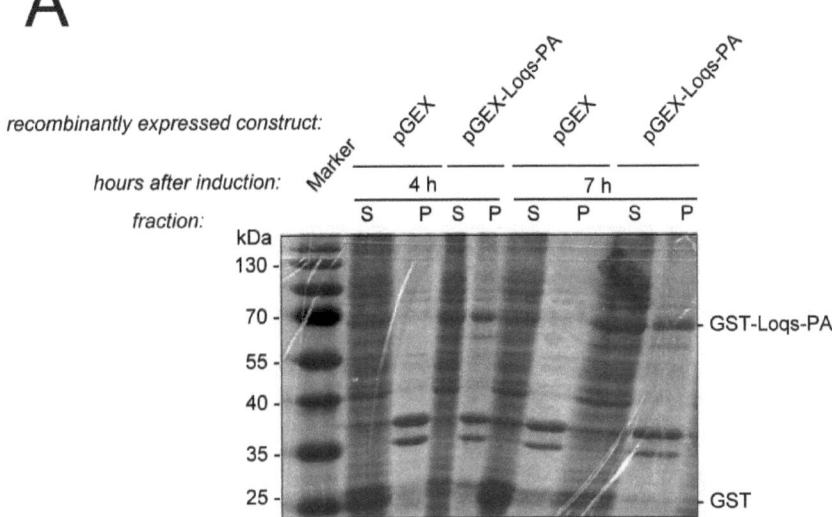

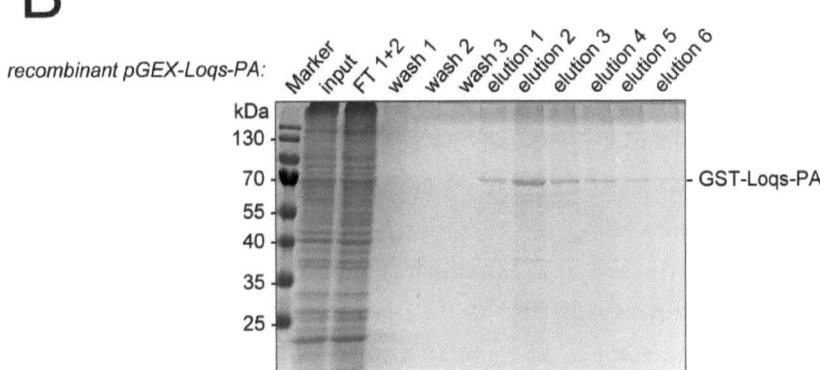

Figure 23: Recombinant expression of GST-Loqs-PA and affinity purification
A) Recombinant expression of pGEX-Loqs-PA 4h and 7h after induction with 0.1 mM IPTG in BL21 *E. coli* cells; empty pGEX-6P-1 served as a control; 50 µl of soluble fraction (S) and an equivalent amount of insoluble pellet (P), boiled in Laemmli SDS loading dye, were separated on a 10% polyacrylamide gel. Gel was stained with colloidal Coomassie.
B) Affinity purification of recombinant GST-Loqs-PA over a Glutathione Sepharose column. 10 µl of total protein extract (input), flow through (FT 1+2) and wash fractions (wash 1-3) and 22.5 µl of elution fractions (E 1-6) were loaded onto a 10% polyacrylamide gel. Gel was stained with colloidal Coomassie.

6 Discussion

The results presented in this thesis extend and refine our understanding of a small-RNA mediated host defense system against selfish genetic elements in somatic cells. I have identified a novel isoform of the Dicer-partner Loquacious, thus explaining how endo-siRNA and miRNA biogenesis pathways are kept distinct. I could demonstrate in this study that interaction of the PD-specific amino acids of Loqs-PD with the N-terminal helicase domain of Dcr-2 is essential for the biogenesis of long hairpin- and high-copy transgene-derived endo-siRNAs. I could further show that R2D2 acts as an antagonist of Loqs-PD in endo-siRNA biogenesis and Loqs-PD antagonizes R2D2 in exo-siRNA mediated silencing. My results indicate that the relative level of Loqs-PD in comparison to the other Loqs isoforms is critical for optimal efficiency of endo-siRNA biogenesis and function. Furthermore, the establishment of a GFP-based reporter system for endo-siRNA silencing demonstrates that the endo-siRNA response can be initiated *de novo*.

6.1 Distinct Loqs isoforms separate the biogenesis routes for endo-siRNAs and miRNAs

The discovery of a new Loqs protein isoform as a central player in endo-siRNA biogenesis sheds light onto the question of how a single gene - *loqs* - can participate in the recognition of pre-miRNA structures together with Dcr-1 (Forstemann et al., 2005; Jiang et al., 2005; Saito et al., 2005) and dsRNA together with Dcr-2 (Czech et al., 2008; Okamura et al., 2008b). The isoform-specific knock-down experiments in this thesis demonstrate that all three currently known *Drosophila* small RNA biogenesis pathways in somatic cells are defined by a distinct set of required factors and can be genetically separated: Dcr-1 and Loqs-PB for miRNAs (including miRtrons; see Figure 1), R2D2 for siRNAs – even those derived from transgenic hairpin constructs (Forstemann et al., 2005) – and Loqs-PD for endo-siRNAs (Figure 24A). It should be noted, however, that this exclusively linear model is to some extent an oversimplification. Certain miRNAs can become incorporated into Ago2-complexes (Förstemann, 2007; Seitz et al., 2008), Loqs and Dcr-1 play a minor but detectable role in transgenic RNAi (Lee et al., 2004; Forstemann et al., 2005) and endo-siRNAs can persist to a certain extent in the absence of *ago2* or *loqs* (Czech et al., 2008; Okamura et al., 2008b).

The current results are, however, insufficient to decide whether Loqs-PD is required for the processing of endo-siRNA precursors or for the Ago2-loading step, or for both. From the perspective of the Loqs protein isoforms, we now know that Loqs-PB participates in miRNA biogenesis and Loqs-PD participates in endo-siRNA biogenesis. If isoform-specific regulation of small RNA silencing pathways proves to be a general mechanism, this will leave the open question for the substrates of the Loqs-PA isoform.

6.2 Antagonism of small RNA biogenesis pathways

All studies concerned with endo-siRNA silencing in *Drosophila* unanimously found a dependence on Ago2 and Dcr-2 (Chung et al., 2008; Czech et al., 2008; Ghildiyal et al., 2008; Kawamura et al., 2008; Zhou et al., 2009). The surprising finding was the requirement of Loqs instead of the canonical RNAi factor R2D2. R2D2 had no influence on the biogenesis of endo-siRNAs and could not rescue long hairpin-derived endo-siRNA biogenesis. Yet, a recent study reported impaired endo-siRNA silencing in *r2d2* mutant fly tissue (Marques et al., 2010). This was attributed to defective loading of endo-siRNAs by a common RISC loading complex (RLC) for endo- and exo-siRNAs consisting of Dcr-2 and R2D2. In addition, they proposed that Loqs-PD was involved not only in endo-siRNA biogenesis but also in exo-siRNA biogenesis. My results suggest a major difference between flies and S2 cells. This is most obvious in the GFP-based endo-siRNA reporter system, which reacts with hyper-repression to depletion of R2D2 alone or in combination with Loqs isoforms, demonstrating the inhibitory effect of R2D2 on endo-siRNA silencing in S2 cells. My RNAi reporter assay, on the other hand, shows an antagonistic effect of Loqs-PD on siRNA-mediated silencing. These results are in direct contrast to the previous study in mutant flies. However, these differences between cell culture and whole animals do not necessarily contradict each other. S2 cells are an immortalized embryonic cell line (Schneider, 1972) and it is therefore not surprising that some of their properties can differ from those of living animals. On the other hand, they are a useful tool to analyze endo-siRNA responses: There is an enrichment of transposable elements in their genome (Finnegan et al., 1978; Potter et al., 1979; Di Franco et al., 1992; Maisonhaute et al., 2007) that probably requires a strong endo-siRNA response to maintain genomic stability. One could speculate that this causes the high level of Loqs-PD expression in S2 cells, while Loqs-PD levels in adult flies are much lower and vary in different tissues (Forstemann et al., 2005; Miyoshi et al., 2010).

This discrepancy between results in flies and in cell culture points out the need for further *in vivo* studies of the endo-siRNA system. This is especially important since all experiments in flies have so far indiscriminately depleted all isoforms, which may account in part for the difference to Loqs-PD-specific RNAi in cell culture. Generation of flies expressing only the Loqs-PD isoform will allow detailed functional and biochemical studies of effects on adult flies. However, lack of miRNA regulation due to loss of Loqs-PB in Loqs-PD rescue flies severely impairs embryogenesis and causes sterility of surviving offspring, similar to the Loqs^{KO} fly (Park et al., 2007). But since overexpression of Loqs-PD in flies with wildtype *loqs* background enhances endo-siRNA biogenesis, a functional distinction between Loqs-PB dependent miRNA biogenesis and Loqs-PD dependent endo-siRNA biogenesis in the adult *Drosophila* fly is already obvious.

My thesis demonstrates that endo- and exo-siRNA pathways in *Drosophila* compete with each other for Dcr-2 through their specific dsRBPs, Loqs-PD and R2D2, respectively (

Figure 24B). My co-immunoprecipitation experiments indicate that there is only a minimal amount of Loqs-PD associated with R2D2. Given the much stronger interaction of Loqs-PD with other Loqs isoforms this is likely to represent an overexpression artifact rather than a ternary complex of Loqs-PD, R2D2 and Dcr-2 as reported before (Miyoshi et al., 2010). My co-immunoprecipitation data for truncated Dcr-2 in addition with published data for interaction of R2D2 with the helicase domain of Dcr-2 (Ye et al., 2007; Lim do et al., 2008) rather suggest that Loqs-PD and R2D2 compete for Dcr-2 binding, for example via a shared interaction site located in the N-terminal DExH/D helicase domain. A parallel phenomenon has been reported for competition of Loqs-PA and -PB in miRNA biogenesis (Liu et al., 2007). There, the authors proposed that the two isoforms compete for binding to Dcr-1 and induce different conformations (Liu et al., 2007). Possibly the Loqs-PB induced conformation increases Dcr-1 affinity for miRNA precursors thus enhancing dicing efficiency (Jiang et al., 2005). One could speculate that Loqs-PD and R2D2 may similarly favor different conformations of Dcr-2, enabling the protein to preferentially act in either endo- or exo-siRNA silencing.

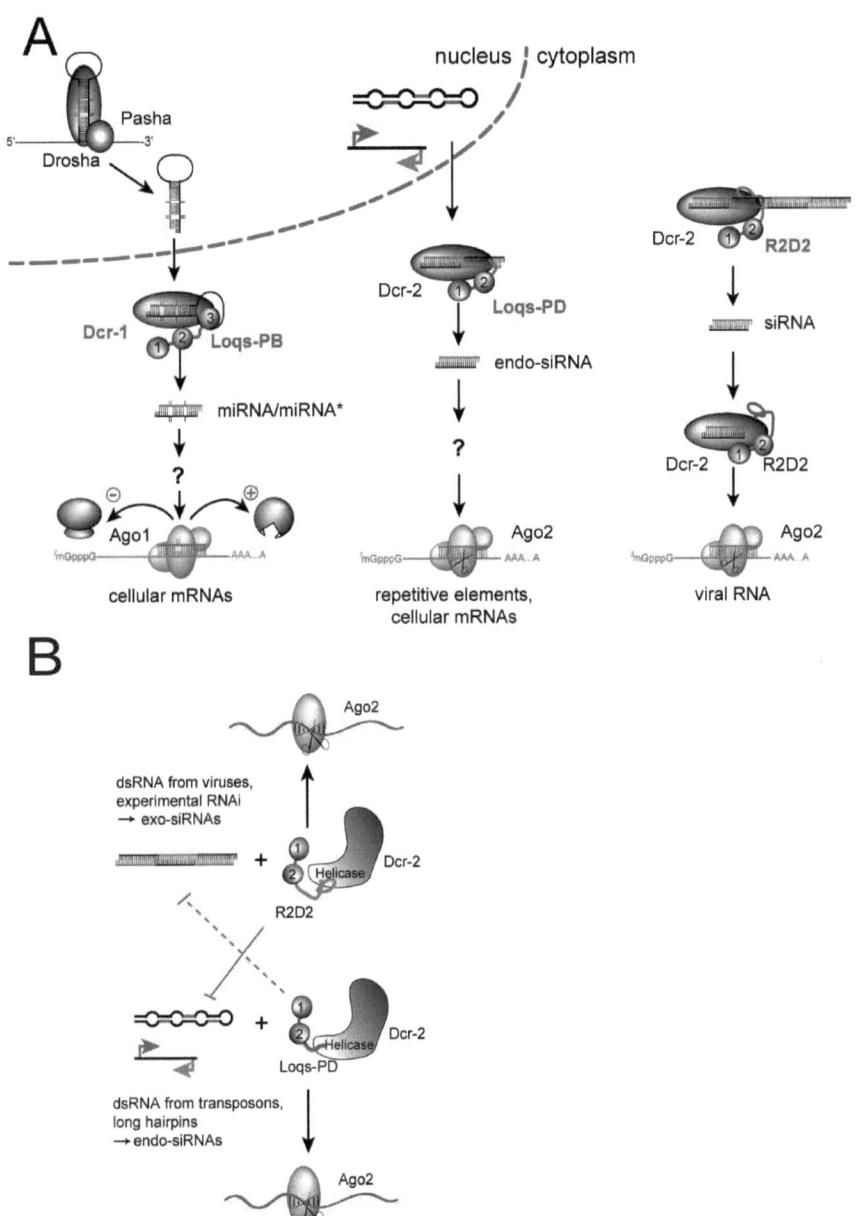

Figure 24: Small RNA biogenesis pathways (legend continued on p. 109)

(legend Figure 24 continued)
A) All somatic small RNA biogenesis pathways can be distinguished by at least one specific component. The model represents important biogenesis steps for miRNAs, endo-siRNAs and siRNAs. Pathway-specific components are labeled in red. For the Loqs isoforms and R2D2, the individual dsRBDs are numbered to make the differences more obvious. While a RISC-loading complex (RLC) has been discovered for siRNAs (Pham et al., 2004; Tomari et al., 2004), analogous complexes for endo-siRNAs and miRNAs have yet to be identified (indicated by question marks).
B) Loqs-PD and R2D2 both bind to the N-terminal helicase domain of Dcr-2, analogous to the situation of human Dicer and TRBP/PACT. Despite the similar architecture, the two *Drosophila* complexes are functionally distinct: My results support the notion that the Dcr-2/R2D2 complex serves as an antagonist of endo-siRNA biogenesis and *vice versa*.

In *C. elegans* the specialized ERI endo-siRNA pathway was defined by mutations that enhance exo-siRNA silencing (Kennedy et al., 2004; Duchaine et al., 2006). This phenomenon was similarly attributed to relaxed competition for limiting common silencing components (Duchaine et al., 2006; Yigit et al., 2006).

The protein partners of human Dicer, TRBP and PACT, are related dsRBD proteins (Patel et al., 1998; Duarte et al., 2000) and – like Loqs-PD and R2D2 – have antagonistic effects on Dicer: TRBP stimulates miRNA dicing and stabilizes Dicer, while PACT inhibits miRNA processing (Ma et al., 2008). In addition to their function in RNAi, the two proteins are involved in the antagonistic regulation of the dsRNA-activated Protein Kinase (PKR). Whereas TRBP inhibits PKR function both by direct dsRBD-mediated binding of PKR and sequestering of dsRNA (Park et al., 1994; Daher et al., 2001), PACT binds to PKR through its dsRBDs 1 and 2 and activates the enzyme via the C-terminus (Huang et al., 2002; Gupta et al., 2003). Interestingly, TRBP interacts with the same residues of PKR but its C-terminal dsRBD has an inhibitory effect (Gupta et al., 2003). Thus, the modulation of enzyme activity by alternative dsRBD protein partners may be a conserved strategy even between otherwise unrelated protein complexes.

6.3 Competition between Loqs isoforms

I have demonstrated in this study that overexpression of isoforms not involved in endo-siRNA silencing inhibits endo-siRNA biogenesis and subsequently impairs silencing of the reporter GFP-transgene. In addition, I could show that depletion of all other Loqs isoforms except for Loqs-PD enhances the efficiency of endo-siRNA mediated silencing, indicating that competition also exists at endogenous levels. The mechanistic basis for this competition may be the sequestering of Dcr-2 via low-affinity binding by Loqs-PB/-PA. This hypothesis is supported by co-immunoprecipitation experiments that indicated a possible interaction of overexpressed Loqs-PA and -PB with Dcr-2. My thesis demonstrates that the Loqs-PD-specific sequence can be sufficient for Dcr-2 binding and is essential for endo-siRNA production. Secondary interactions of the first two dsRBDs of Loqs with either an RNA substrate or Dcr-2 directly may also contribute to binding and thus explain low-affinity interactions between Dcr-2 and Loqs-PB/-PA. A similar competition phenomenon has been observed for Dcr-1: Both Loqs-PB and Loqs-PA readily interact with Dcr-1 *in vitro*, but when presented together Dcr-1 prefers binding to Loqs-PB (Ye et al., 2007). Interestingly, the 46 additional amino acids in Loqs-PB are placed at the same position within Loqs (counting from the N-terminus) as the 22 amino acids specific to Loqs-PD (Table 8). Loqs-PA and Loqs-PC are similarly characterized by distinct sequences at this position.

Table 8: Isoform specific amino acid sequences situated at the same position from the N-terminus of the protein
22 amino acid Loqs-PD-specific sequence and 46 amino acid sequence lacking in Loqs-PA are indicated; in the Loqs-PA isoform only the common C-terminus encompassing the third dsRBD can be found; the Loqs-PC-specific sequence is listed for completeness, even though endogenous expression cannot be verified on protein level; "-" marks stop codon in the corresponding mRNA

Loqs-PD isoform: Loqs-PDspec (22 amino acids)

VSIIQDIDRYEQVSKDFEFIKI-

Loqs-PB isoform: Sequence shared by Loqs-PB and -PC but not -PA (46 amino acids)

PRSSENYYGELKDISVPTLTTQHSNKVSQFHKTLKNATGKKLLKLQ

Loqs-PA isoform: Common C-terminus of Loqs-PB and Loqs-PA (including 3rd dsRBD) (82 amino acids)

KTCLKNNKIDYIKLLGEIATENQFEVTYVDIEEKTFSGQFQCLVQLSTLPVGVCHGSGPTAADAQRHAAQNALEYLKIMTKK-

Loqs-PC isoform: Loqs-PCspec (43 amino acids)

NESVKHLFHTKVICFNSPLACVISNVCEMQWRKETKNFALLFT

Since different dsRBDs within one protein can have distinct functions (reviewed in Doyle et al., 2002), it will be necessary to attribute roles to the first two dsRBDs in both Loqs and

R2D2 as well. Interestingly, all Loqs isoforms share the first two dsRBDs. This suggests that despite the involvement of Loqs isoforms in different small RNA silencing pathways they may require conserved functions. However, one could speculate that a different Dicer-partner could stimulate a conformational change in the double-stranded RNA binding domain protein, which would increase versatility. NMR studies of dsRBP/Dicer interaction will help in this analysis of individual domains complemented by a functional approach based on my domain-swapped dsRBD chimaeras. By transfecting different small RNA reporter systems with different combinations of Loqs and R2D2 dsRBDs it will be possible to elucidate conserved and unique roles of individual domains.

The ratio of individual Loqs isoforms is not uniform in different fly tissues or cell culture cells (Forstemann et al., 2005; Miyoshi et al., 2010). This suggests that their expression can be regulated individually. Since Loqs-PB is essential for the miRNA pathway and Loqs-PD has to respond to the threat of transposons, it seems likely that the endo-siRNA factor could be regulated by a separate mechanism. Loqs-PD not only has a unique C-terminus but also a 3′UTR that is not shared by any of the other isoforms. RNAi against this alternatively poly-adenylated transcript mirrors a possible endogenous mechanism to regulate the stability of *loqs*-RD independently from the other isoforms. Since the 3′UTR of an mRNA can contain possible binding sites for regulatory factors (Shalgi et al., 2005; Xie et al., 2005), the cell would be able to adapt the abundance of Loqs isoforms for miRNA and endo-siRNA silencing independently and according to the cell's condition.

6.4 A conserved interaction scheme between Dicer and dsRBD proteins

The single Dicer enzyme found in humans can interact with TRBP and PACT, both of which are dsRBD proteins and homologs of *Drosophila loqs* and *r2d2* (Gatignol et al., 1991; Chendrimada et al., 2005; Haase et al., 2005; Lee et al., 2006). This interaction could be mapped to 69 amino acids at the C-terminal end of TRBP and 165 amino acids in the N-terminal domain of Dicer (Daniels et al., 2009). Thus, the positions of the interacting domains are conserved not only between *Drosophila* Dcr-1 and Dcr-2, but also in vertebrate Dicer enzymes. In *C. elegans*, RDE-4, interacts with Dcr-1 as well as another helicase domain containing protein, Drh-1 (Tabara et al., 2002). While the Dicer-interacting domain of RDE-4 has not been mapped, the C-terminal third dsRBD mediates dimerization of RDE-4. The dimer binds long double-stranded RNA cooperatively and shows a length-dependent decrease in binding affinity between a 650 nt fragment of dsRNA and a mature siRNA *in vitro*. This decrease is exacerbated by deletion of the C-terminus, starting at the end of the second dsRBD (Parker et al., 2008). Truncated RDE-4 does not form homodimers any more and cannot sustain the production of siRNAs *in vitro* (Parker et al., 2006).

In this study I report the formation of homomultimers between Loqs-PB as well as heteromultimers between Loqs-PB and Loqs-PA. Heteromultimers between Loqs-PD and Loqs-PB were much less abundant or undetectable at endogenous levels. This finding is consistent with a conserved role of the third dsRBD in mediating dimerization of the dsRBD proteins. Whether such a dimerization is possible while retaining the association with Dicer or whether the dimers only form in the population of free dsRBD proteins remains to be elucidated. In the context of PKR activation, TRBP and PACT form an inactive heterodimer which cannot stimulate PKR activity (Laraki et al., 2008; Daher et al., 2009). Thus, the multimerization may serve a regulatory purpose. It is interesting to note that there is a splice variant of human TRBP lacking the third dsRBD, thus resembling Loqs-PD (Haase et al., 2005). Does this also represent a functionally distinct isoform with a modified protein-protein interaction profile?

Deletion analysis indicated that both the 22 amino acid Loqs-PD-specific sequence and the Dcr-2 helicase domain are necessary for endo-siRNA function. However, I could not observe any impairment of miR-277 perfect match reporter silencing, when I transfected the reporter with a truncated Flag-Dcr-2 construct, lacking the helicase domain. Since miR-277 is diced by Loqs-PB/Dcr-1 and then loaded into Ago2 by the Dcr-2/R2D2 RLC, this suggests that the helicase domain is not required for Dcr-2/R2D2-dependent loading. This is consistent with previously published data about the uncoupling of the dicing and the loading step in dcr-2^{G31R} mutant flies. This allele carries a point mutation that abolishes the nuclease function of Dcr-2 (Förstemann, 2007) but leaves the loading capacity unimpaired. The deletion of the helicase region may therefore similarly not interfere with Dcr-2-dependent loading of miR-277 into Ago2.

Another conserved theme in dsRBP/Dicer interaction is a possible mutual stabilization between the binding partners (Liu et al., 2003; Forstemann et al., 2005; Förstemann, 2007;

Liu et al., 2007). Especially destabilization of R2D2 after Dcr-2 depletion is pronounced (Liu et al., 2003; Förstemann, 2007) and may contribute to maintaining a balance between competing pathways (see next paragraph).

6.5 Affinity *versus* abundance: Stability of the siRNA system and a model for siRNA precursor recognition

My results suggest a direct correlation between the relative level of Loqs-PD compared to competing dsRBPs and the efficiency of the endo-siRNA response. The competition between Loqs isoforms is directly obvious from the isoform-specific RNAi experiments in the 63N1 reporter system: Loqs-PD-specific depletion has a more pronounced effect on the reporter than depletion of all isoforms. The RNAi trigger against the 5´UTR of *loqs* is less efficient in depleting the Loqs-PD isoform than the *loqs-ORF* RNAi construct. This is directly mirrored by the reporter response. The flexibility of the system is immediately apparent in my co-immunoprecipitation data as well, where overexpression can induce less favorable interactions between Dicer proteins and Loqs isoforms. This suggests that affinities and concentrations of binding partners are in equilibrium under steady-state conditions in the cell. Furthermore I could show that R2D2 depletion in the endo-siRNA reporter causes hyper-repression of the GFP level and double knock-down experiments suggested that the ratio of R2D2 and Loqs-PD is essential for the balance between endo- and exo-siRNA pathways. This equilibrium, however, is precarious given that siRNAs from endogenous loci are far less abundant, namely by approximately two orders of magnitude (Okamura et al., 2008a). When a cell is faced with a viral infection, it is flooded with double-stranded RNA of viral origin. There are two possible measures to prevent the siRNA system of a cell from saturation and to maintain transposon surveillance even during viral infections: First, Loqs-PD affinity for Dcr-2 should be higher than R2D2 affinity for Dcr-2. This would avoid the complete sequestering of Dcr-2 by exo-siRNA factors. Second, since the substrate of the endo-siRNA pathway, namely long double-stranded RNA derived from endogenous loci, does not differ in structure from double-stranded RNA from viral origin, the Dcr-2/Loqs-PD and the Dcr-2/R2D2 complexes should have distinct chemical properties. A possible hypothesis would be that the Dcr-2/R2D2 heterodimer has a lower affinity that requires a higher abundance of double-stranded RNA substrate, while the Dcr2/Loqs-PD complex has a high affinity allowing substrate binding at lower double-stranded RNA concentrations (K.

Förstemann, personal communication). Recombinant expression of Dicer and dsRBPs as well as Loqs/R2D2-chimaeras will allow measuring of binding constants between pathway components. Additionally, affinity for RNA substrate of individual proteins can be determined *in vitro*. Taken together, recombinantly expressed proteins may aid to attribute function to individual domains and help understand pathway specificity of small RNAs.

6.6 Comparison between two reporter cell lines: 63N1 and 63-6

Transient transfection of cell culture cells has several drawbacks: First, transfection reagents are usually harmful for a cell. Second, expression levels of individual cells vary widely depending on the number of plasmids in the cytoplasm. Third, expression declines soon after transfection as the number of plasmids in a single cell decreases during proliferation. Formation of clonal cell lines stably expressing a transgene has the advantage that the entire population of cells has a uniform expression level. However, the number of transgenes integrated in the genome can vary between individual cell lines. Both the 63-6 and the 63N1 cell lines were derived from transfection of S2 cells with a GFP-expression plasmid, however from different single cell clones. While the 63N1 cell line shows a strong response to depletion of endo-siRNA factors like Ago2, Dcr-2 and Loqs, the increase in GFP levels of the 63-6 cell line is much less pronounced. The most straightforward explanation for this is a difference in the copy number of integrated GFP-genes in the genome of the two lines. Quantitative PCR of genomic DNA performed by a fellow PhD-student in the laboratory, Stephanie Esslinger, indeed showed an approximately 40-times higher copy number in the 63N1 genome than in the 63-6 genome (Hartig et al., 2009). The more efficient endo-siRNA response to the higher number of transgenes implies that the cell can sense the copy-number of repetitive genetic elements. A possible explanation for this sensitivity is the pervasive transcription of intergenic regions in *Drosophila* and other species (reviewed in Kapranov et al., 2007). Possible transcription events in antisense orientation would produce complementary strands for mRNA derived from selfish genetic elements or transgenes and allow processing of endo-siRNAs from these loci. This hypothesis would explain the different amplitudes of reporter reaction in the 63N1 and the 63-6 cell lines based on different copy-numbers of GFP transgenes integrated in the genome.

However, there must be additional genetic differences between the two cell lines. This is obvious by the distinct reactions of the 63N1 and the 63-6 cell lines to depletion of p68/Lip. What causes this pronounced de-repression of GFP levels in the 63-6 line and whether there is a transcriptional component involved in silencing p68/Lip in 63-6 cells, remains unclear (see next paragraph).

6.7 Transcriptional *versus* post-transcriptional gene silencing

The 63N1 endo-siRNA cell culture reporter cell line did not show any significant response to depletion of putative transcriptional silencing factors. FMR1, p68/Lip and HP-1 are involved in chromatin remodeling and silencing of gene expression by tight packaging of DNA into heterochromatin. This suggests that the endo-siRNA response is primarily caused by post-transcriptional cleavage of the target. This is consistent with the efficient endonuclease activity of Ago2 (Förstemann, 2007). Post-transcriptional target cleavage has indeed been reported before *in vitro* (Kawamura et al., 2008; Okamura et al., 2008b) and *in vivo* (Czech et al., 2008). With the help of a nascent RNA labeling and purification strategy a fellow PhD student in our laboratory, Stephanie Esslinger, could show that endo-siRNAs directed against a transgene construct predominantly affect the degradation rate of the corresponding transcripts and do not impose transcriptional regulation. Taken together our findings suggest that a post-transcriptional mechanism is sufficient to explain most – if not all – of the repression mediated by endo-siRNAs.

Compacted heterochromatin, on the other hand, could also have a global effect on the rate of transcriptional noise and pervasive transcription. Depletion of heterochromatin factors could cause decondensation of DNA thus increasing the level of possible antisense transcripts. This in turn would lead to a more pronounced endo-siRNA response. However, I could not detect this effect using the endo-siRNA cell culture reporter.

6.8 Is there an RdRP-like activity in Drosophila?

The RNA-dependent RNA polymerase (RdRP)-mediated generation of secondary siRNAs in the yeast *S. pombe*, in *C. elegans*, and in plants increases and elongates the RNAi response (reviewed in Ghildiyal et al., 2009). In principle, target mRNA serves as a template for the

RNA polymerase. Secondary siRNAs can then in turn be loaded into RISC complexes. Since the secondary siRNAs are not identical with the primary siRNA but templated by novel target sequence, the silencing effectively "spreads" along the genome (Sijen et al., 2001). This phenomenon makes isoform-specific RNAi in *C. elegans* difficult or impossible. No obvious RdRP activity could be found in flies or mammals (Zamore et al., 2000; Nykanen et al., 2001; Martinez et al., 2002; Haley et al., 2003; Roignant et al., 2003; Tomari et al., 2004). In addition, genome analysis found no apparent gene encoding for a canonical RdRP in *Drosophila*. Deep-sequencing data showed no endo-siRNA sequences crossing an exon-exon junction, which is inconsistent with a potential RdRP activity (Hartig et al., 2009). However, an RdRP-like activity has been described in *Drosophila*, essentially by the group of Bruce Paterson (Lipardi et al., 2001). Recently, the same group biochemically attributed the non-canonical RdRP-activity to D-elp-1, a subunit of the PolII elongator complex (Lipardi et al., 2009). My GFP-based reporter system for siRNA silencing was analogous to the one used in the Paterson group, but I could not corroborate the RdRP-like activity of D-elp-1. If the D-elp-1 showed similar properties as RdRPs in *C. elegans*, one would suppose that isoform-specific RNAi – as employed in this thesis and reported before (Roignant et al., 2003) – would not be possible. The majority of the evidence thus indicates that an RdRP-like activity does only play a minor role in *Drosophila*, if at all.

7 Abbreviations

+	in chimaeric constructs: ligation-dependent addition of two codons (glycine and lysine)
µg	microgram
63N1	endo-siRNA cell culture reporter cell line
Ago	Argonaute protein
Amp	ampicillin
APS	ammonium peroxodisulfate
ATP	adenosine triphosphate
bd.	bound IP fraction
BLAST	Basic Local Alignment Search Tool
bp	base pair(s)
BSA	bovine serum albumine
cDNA	complementary DNA
CG4068	hairpin forming endo-siRNA precursor gene
co-IP	co-immunoprecipitation
C-term	protein C-terminus
CT-value	cycle of threshold value in qPCR
d	day(s)
D. melanogaster	*Drosophila melanogaster*
dcr	dicer gene
Dcr	Dicer protein
DExH/D	superfamily II helicase family member that works on RNA
DMSO	dimethyl sulfoxide
DNA	desoxy-ribonucleic acid
dNTP	desoxy-nucleotide-tri-phosphate
ds	double-stranded
dsRBD	double-stranded RNA binding domain
dsRBP	double-stranded RNA binding domain protein
dsRNA	double-stranded RNA
DTT	dithiothreitol

DUF	Domain of Unknown Function
ECL	Enhanced Chemiluminescence
E. coli	*Escherichia coli*
EGFP	Enhanced Green Fluorescent protein
EM	electron microscopy
endo-	endogenous
endo-siRNA	endogenous small interfering RNA
exo-	exogenous
exo-siRNA	exogenous small-interfering RNA
FACS	Fluorescence Activated Cell Sorting
FBS	Fetal Bovine Serum
FMR1	Fragile Mental Retardation Protein 1
GFP	Green Fluorescent Protein
GST	glutathione S-transferase
h	hour(s)
hel	helicase domain
HP-1	Heterochomatin Protein 1
HRP	Horseradish Peroxidase
Hygro	hygromycin
IgG	immunoglobulin protein G
inp.	input
IP	immunoprecipitation
IPTG	Isopropyl-β-D-thiogalactopyranosid
Kana	kanamycin
KLH	Keyhole limpet hemocyanin
KO	knock-out
L	double-stranded RNA binding domain of Loquacious
l.c.	antibody light-chain
loqs	loquacious gene
Loqs	Loquacious protein
Luc	luciferase
MCS	Multiple Cloning Site

mg	milligram
miR	micro RNA
miRNA	micro RNA
ml	milliliter
mRNA	messenger RNA
Neo	neomycin
ng	nanogram
nt	nucleotide(s)
N-term	protein N-terminus
NTP	nucleotide-tri-phosphate
OD	optical density
ORF	open reading frame
OrR	Oregon R fly wildtype strain
p.a.	pro analysi
PA/PB/PC/PD	protein isoform A/B/C/D
PACT	protein activator of interferon-induced double-stranded RNA-dependent protein kinase
PAGE	Polyacrylamide Gel Electrophoresis
PAZ	Piwi-Argonaute-Zwille domain of Dicer and Argonaute proteins
PKR	interferon-induced double-stranded RNA-dependent protein kinase
PCR	Polymerase Chain Reaction
PCspec	Loqs-PC-specific sequence
PDspec	Loqs-PD-specific sequence
piRNA	Piwi-interacting RNA
PNK	polynucleotide kinase
Pol II	DNA polymerase II
Poly-A	poly-adenylation
PVDF	Polyvinylidenfluoride
qPCR	quantitative Polymerase Chain Reaction
R	double-stranded RNA binding domain of R2D2
R2D2	protein name derived from: 2 dsRBD-containing protein interacting with Dcr-2

RA/RB/RC/RD	*loqs* isoform A/B/C/D mRNA
RACE	rapid amplification of cDNA ends
rb	rabbit
RdRP	RNA-dependent RNA-Polymerase
rel.	relative
RISC	RNA induced silencing complex
RLC	RISC loading complex
RNA	ribonucleic acid
RNAi	RNA interference
RNaseIII	endoribonuclease class III
rRNA	ribosomal RNA
RT	reverse transcription or real-time
S. pombe	*Schizosaccharomyces pombe*
S2 cell	Schneider-2 cell
SD	standard deviation
siRNA	small interfering RNA
SOB	Super Optimal Broth
SSC	sodium chloride/sodium citrate
SV40	Simian Virus 40
T	thymin
TAR	transactivation response RNA
tech.	technical
TRBP	TAR RNA binding protein
tub.	tubulin
U	uracil
UAS	yeast Upstream Activating Sequence
UTR	untranslated region
V	Volt
Δ	deletion
α	anti
°C	degrees Celsius
μ	micro

8 Appendix

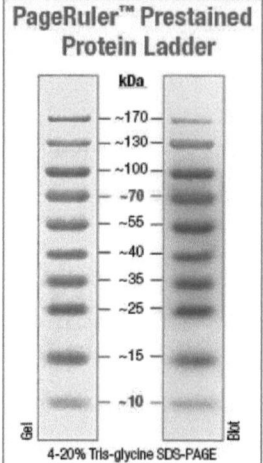

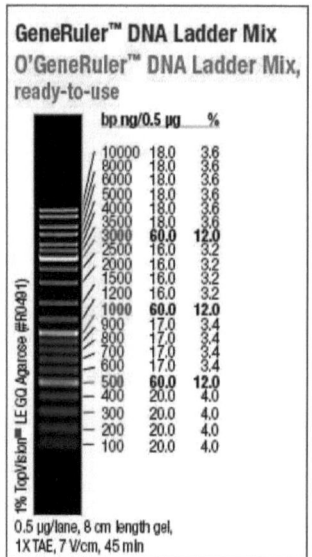

Appendix 1: Protein and DNA markers
A) PageRuler Prestained Protein Marker (Fermentas)
B) Gene Ruler DNA Ladder Mix (Fermentas)

Appendix

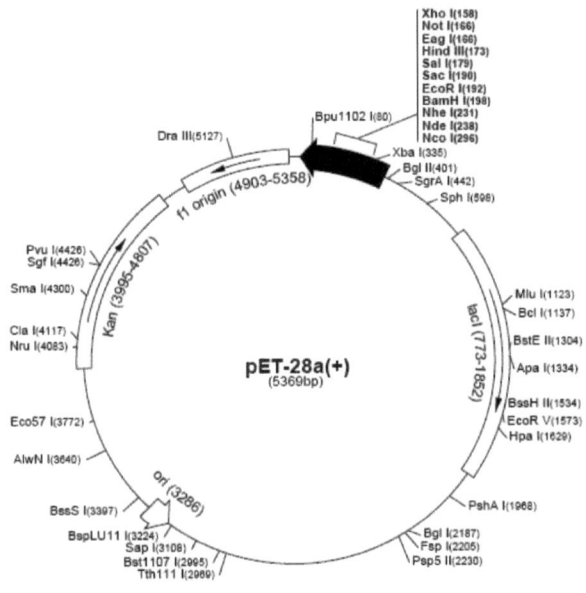

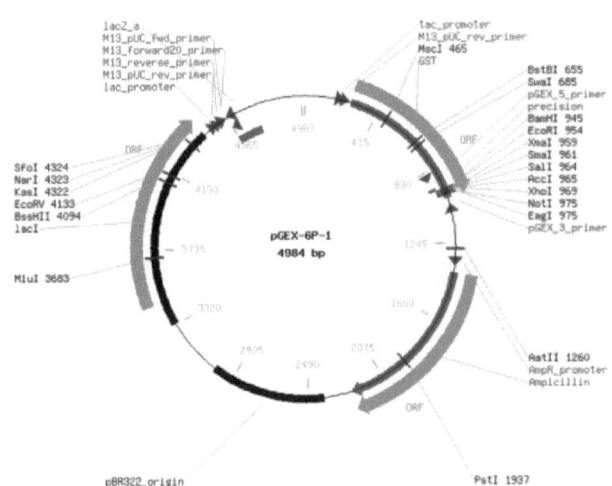

Appendix 2: Vectors for recombinant expression
A) Plasmid map for pET-28a (Novagen)
B) Plasmid map for pGex-6P-1 (Amersham)

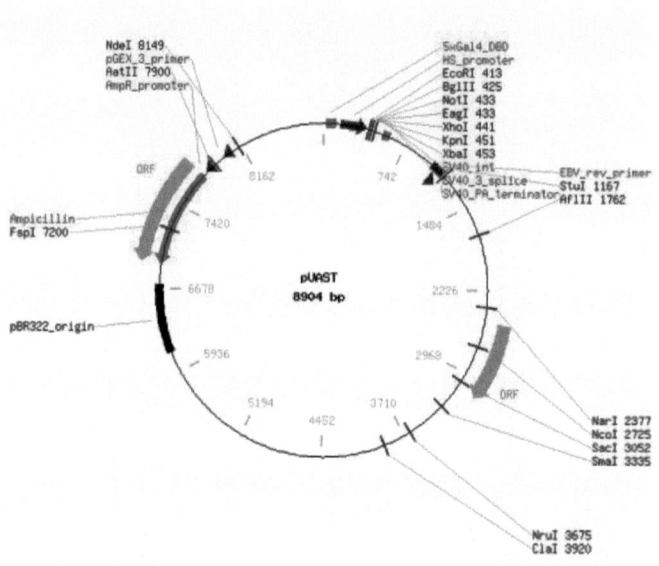

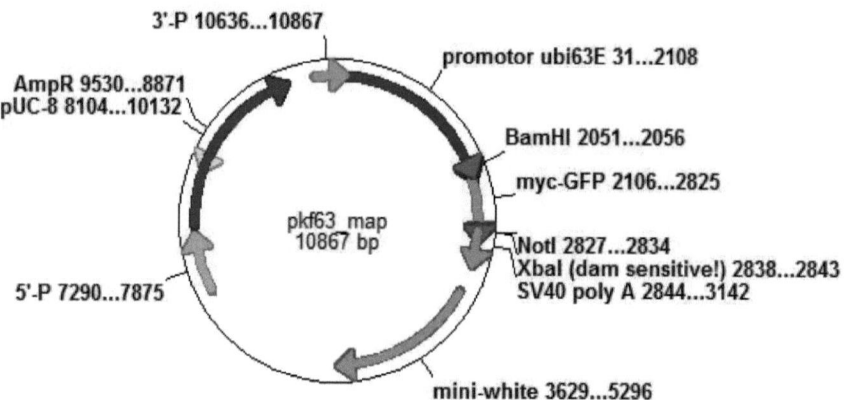

Appendix 3: Expression vectors
A) Plasmid map for pUAST; conditional UAS/Gal4 expression system
B) Plasmid map for pKF63; myc-GFP expression under tubulin promotor

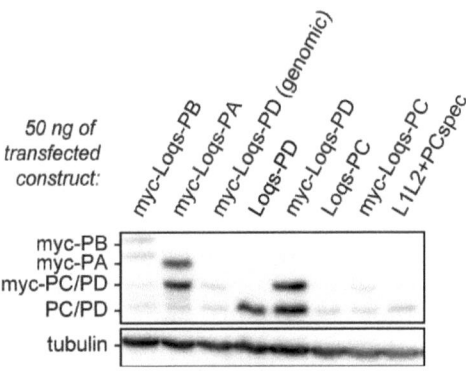

Appendix 4: Loqs-PC expression
Western blot from S2 cells transfected with 50 ng of various Loqs expression constructs. If the same amount of construct was transfected, I observed varying amounts of expression. For all experiments I adapted plasmid amounts for equal expression levels, however, I was unable to achieve high expression for Loqs-PC constructs.

Appendix 5: Table for Poly-A prediction results

Nucleotide positions refer to genomic region of loqs (http://flybase.org/reports/FBgn0032515.html); *loqs*-RD 3´UTR from position 1445-1789; yellow highlighted sites mark potential poly-A sites for *loqs*-RD; a prediction algorithm for human poly-A sites was used with standard parameters (Ahmed et al., 2009)

Start position	Sequence	Score	Prediction
1913	GTGATA	169.439	Positive
1676	GTACTA	137.023	Positive
1933	ATTGGC	133.975	Positive
1816	TTATAT	104.823	Positive
1874	TTAAAG	0.99965	Positive
1948	TTCAAA	0.99769	Positive
1692	TTTATA	0.95536	Positive
1860	GCATGG	0.92238	Positive
1607	CGTTTA	0.91568	Positive
1663	AGCTAG	0.86657	Positive
1827	TATTTC	0.84898	Positive
1806	TATTTA	0.83740	Positive
2875	AGACAA	0.83251	Positive
1842	TTTTCA	0.76249	Positive
1619	TCCTCA	0.68375	Positive
2014	GCAGTA	0.68208	Positive
2860	TACGCT	0.65299	Positive
1959	GCGAAA	0.64682	Positive
1969	ATGGAG	0.58881	Positive
2792	GGCAGT	0.58617	Positive
1565	TAATAC	0.57317	Positive
2259	ATAATA	0.54416	Positive
2785	TTGATT	0.53071	Positive
2774	CACCGA	0.50245	Positive
2279	TGAAAG	0.48552	Positive
1710	ATATTT	0.48217	Positive
1096	CGACAA	0.48173	Positive
539	CAGCCA	0.47180	Positive
1748	TGACAT	0.44816	Positive
1768	TTAACA	0.44018	Positive
783	AGCCGA	0.43998	Positive
1061	ATCCAA	0.43691	Positive
2852	AATGTA	0.41890	Positive
519	CCCAGG	0.40346	Positive
1788	TATTTA	0.39514	Positive
1547	TAAATG	0.37463	Positive
2757	ATATTT	0.35843	Positive
1555	ATATAT	0.34197	Positive
2815	TATGTA	0.32201	Positive
2742	CATATA	0.29382	Positive
1632	ATATCC	0.29366	Positive
527	GCGCAG	0.29118	Positive
2749	ATATAT	0.28689	Positive
2805	GTTAAC	0.27847	Positive

```
ATGGACCAGGAGAATTTCCACGGCTCCAGCTTGCCGCAGCAGCTACAGAACCTCCACA
TCCAGCCGCAGCAGGCGTCCCCCAATCCTGTCCAGACGGGATTTGCTCCACGGCGGCA
CTATAATAACCTTGTCGGCCTGGGCAATGGAAATGCCGTCAGTGGTAGTCCGGTGAAG
GGTGCTCCGCTGGGGCAGCGCCATGTGAAGCTCAAGAAGGAGAAGATATCCGCCCAG
GTTGCGCAGCTGTCTCAGCCAGGTCAGCTGCAGCTGTCAGATGTTGGTGATCCTGCCT
TGGCGGGCGGATCGGGCTTACAAGGTGGAGTCGGCCTTATGGGCGTAATATTGCCCAG
CGACGAGGCCTTAAAGTTCGTCAGCGAAACGGACGCCAATGGACTGGCCATGAAGACG
CCCGTCAGCATTCTGCAAGAGCTGCTAAGCCGACGAGGAATTACTCCCGGCTATGAACT
TGTCCAGATCGAGGGCGCCATACATGAGCCGACCTTCCGGTTTCGCGTGTCCTTTAAAG
ACAAGGATACGCCCTTCACGGCCATGGGGGCAGGACGATCGAAGAAGGAGGCCAAGC
ATGCGGCGGCCCGTGCGCTCATCGACAAGCTGATCGGCGCGCAGCTGCCGGAATCGC
CTAGCAGCTCCGCTGGTCCGTCGGTGACTGGGCTCACGGTCGCCGGAAGCGGAGGAG
ACGGCAATGCCAATGCCACAGGCGGAGGAGATGCCAGCGACAAGACCGTTGGTAATCC
GATTGGCTGGTTGCAGGAAATGTGCATGCAACGGCGATGGCCACCGCCGTCGTACGAA
ACGGAAACGGAAGTGGGTCTTCCCCACGAGCGGCTCTTTACGATCGCCTGCTCGATAC
TCAACTACCGCGAGATGGGCAAGGGCAAAAGCAAGAAGATAGCCAAGCGCTTGGCCGC
CCACCGCATGTGGATGCGTCTGCAGGAGACTCCCATCGATTCGGGCAAAATCAGCGAC
AGCATCTGCGGCGAGTTGGAGGGCGAACCCCGCAGTAGTGAAAATTATTATGGTGAATT
GAAAGATATCTCTGTGCCGACACTGACCACGCAGCACAGTAACAAAGTATCCCAGTTCC
ATAAGACCCTAAAAAATGCAACGGGCAAAAAACTGCTTAAGTTACAGAAGACTTGCTTGA
AGAACAACAAGATTGATTACATCAAGCTGCTGGGCGAAATCGCCACGGAGAACCAGTTC
GAGGTGACCTATGTGGACATAGAGGAGAAGACCTTCTCTGGCCAGTTCCAGTGCCTGG
TTCAACTGTCCACGCTGCCCGTTGGCGTTTGCCACGGCAGCGGACCAACAGCTGCCGA
TGCCCAGCGGCATGCCGCCCAGAATGCCCTCGAGTACTTGAAGATCATGACCAAGAAG
TAG
```

MDQENFHGSSLPQQLQNLHIQPQQASPNPVQTGFAPRRHYNNLVGLGNGNAVSGSPVKG
APLGQRHVKLKKEKISAQVAQLSQPGQLQLSDVGDPALAGGSGLQGGVGLMGVILPSDEAL
KFVSETDANGLAMKTPVSILQELLSRRGITPGYELVQIEGAIHEPTFRFRVSFKDKDTPFTAM
GAGRSKKEAKHAAARALIDKLIGAQLPESPSSSAGPSVTGLTVAGSGGDGNANATGGGDAS
DKTVGNPIGWLQEMCMQRRWPPPSYETETEVGLPHERLFTIACSILNYREMGKGKSKKIAK
RLAAHRMWMRLQETPIDSGKISDSICGELEGEPRSSENYYGELKDISVPTLTTQHSNKVSQF
HKTLKNATGKKLLKLQKTCLKNNKIDYIKLLGEIATENQFEVTYVDIEEKTFSGQFQCLVQLST
LPVGVCHGSGPTAADAQRHAAQNALEYLKIMTKK-

Appendix 6: **Nucleotide and amino acid sequence of Loqs-PB**
Sequences corresponding to conserved dsRBD regions are marked (dsRBD1 yellow; dsRBD2 red; dsRBD3 olive);
conserved regions were deduced from BLAST results

```
ATGGATAACAAGTCAGCCGTATCTGCTCTACAGGAGTTTTGTGCCCGGACACAGATTAA
TCTACCAACATACAGTTTTATTCCCGGCGAAGACGGAGGGTACGTCTGTAAAGTTGAAC
TATTGGAGATAGAGGCCCTTGGAAATGGGCGTTCGAAGCGTGATGCCAAACACCTGGC
TGCCAGCAATATCTTGCGTAAAATCCAACTGCTGCCCGGCATACACGGCTTGATGAAGG
ATTCGACTGTGGGTGATCTGGATGAGGAACTGACTAACCTCAACCGGGACATGGTGAA
GGAGCTGCGTGACTACTGCGTCCGCCGCGAGATGCCACTGCCCTGCATTGAGGTAGTG
CAGCAAAGCGGCACCCCGAGCGCCCCGGAATTCGTGGCCTGTTGCTCCGTGGCCTCC
ATAGTACGCTACGGAAAGTCGGACAAAAAGAAGGATGCCCGTCAGCGAGCGGCCATTG
AAATGCTGGCCTTAATCTCCAGCAATTCGGACAATTTGCGTCCGGATCAAATGCAAGTA
GCGAGCACAAGCAAATTGAAAGTTGTTGATATGGAAGAATCTATGGAGGAATTGGAGGC
ATTGCGCAGAAAGAAATTTACCACCTACTGGGAGTTGAAGGAAGCCGGGAGCGTAGAC
CATACAGGCATGCGGCTCTGCGACCGACACAACTACTTCAAGAACTTCTATCCTACCCT
GAAAAAGGAGGCCATTGAGGCCATCAATTCAGATGAATACGAGAGCTCCAAGGATAAG
GCTATGGACGTAATGAGCTCTTTAAAGATAACACCCAAAATCAGTGAAGTGGAATCTTCA
TCGTTGGTTCCCTTGCTTAGCGTCGAGCTTAATTGTGCATTCGACGTGGTCCTTATGGC
AAAGGAGACCGATATCTACGACCATATAATAGACTATTTTCGCACCATGTTGATTTAA
```

MDNKSAVSALQEFCARTQINLPTYSFIPGEDGGYVCKVELLEIEALGNGRSKRDAKHLAASNI
LRKIQLLPGIHGLMKDSTVGDLDEELTNLNRDMVKELRDYCVRREMPLPCIEVVQQSGTPSA
PEFVACCSVASIVRYGKSDKKKDARQRAAIEMLALISSNSDNLRPDQMQVASTSKLKVVDM
EESMEELEALRRKKFTTYWELKEAGSVDHTGMRLCDRHNYFKNFYPTLKKEAIEAINSDEYE
SSKDKAMDVMSSLKITPKISEVESSSLVPLLSVELNCAFDVVLMAKETDIYDHIIDYFRTMLI-

Appendix 7: Nucleotide and amino acid sequence of R2D2
Sequences corresponding to conserved dsRBD regions are marked (dsRBD1 green; dsRBD2 turquoise); conserved regions were deduced from BLAST results

MEDVEIKPRGYQLRLVDHLTKSNGIVYLPTGSGKTFVAILVLKRFSQDFDKPIESGGKRALFM
CNTVELARQQAMAVRRCTNFKVGFYVGEQGVDDWTRGMWSDEIKKNQVLVGTAQVFLDM
VTQTYVALSSLSVVIIDECHHGTGHHPFREFMRLFTIANQTKLPRVVGLTGVLIKGNEITNVAT
KLKELEITYRGNIITVSDTKEMENVMLYATKPTEVMVSFPHQEQVLTVTRLISAEIEKFYVSLD
LMNIGVQPIRRSKSLQCLRDPSKKSFVKQLFNDFLYQMKEYGIYAASIAIISLIVEFDIKRRQAE
TLSVKLMHRTALTLCEKIRHLLVQKLQDMTYDDDDDNVNTEEVIMNFSTPKVQRFLMSLKVS
FADKDPKDICCLVFVERRYTCKCIYGLLLNYIQSTPELRNVLTPQFMVGRNNISPDFESVLER
KWQKSAIQQFRDGNANLMICSSVLEEGIDVQACNHVFILDPVKTFNMYVQSKGRARTTEAK
FVLFTADKEREKTIQQIYQYRKAHNDIAEYLKDRV==LEKTEPELYEIKGHFQDDIDPFTNENGA==
==VLLPNN==ALAILHRYCQTIPTDAFGFVIPWFHVLQEDERDRIFGVSAKGKHVISINMPVNCMLR
DTIYSDPMDNVKTAKISAAFKACKVLYSLGELNERFVPKTLKERVASIADVHFEHWNKYGDS
VTATVNKADKSKDRTYKTECPLEFYDALPRVGEICYAYEIFLEPQFESCEYTEHMYLNLQTP
RNYAILLRNKLPRLAEMPLFSNQGKLHVRVANAPLEVIIQNSEQLELLHQFHGMVFRDILKIW
HPFFVLDRRSKENSYLVVPLILGAGEQKCFDWELMTNFRRLPQSHGSNVQQREQQPAPRP
EDFEGKIVTQWYANYDKPMLVTKVHRELTPLSYMEKNQQDKTYYEFTMSKYGNRIGDVVHK
DKFMIEVRDLTEQLTFYVHNRGKFNAKSKAKMKVILIPELCFNFNFPGDLWLKLIFLPSILNRM
YFLLHAEALRKRFNTYLNLHLLPFNGTDYMPRPLEIDYSLKRNVDPLGNVIPTEDIEEPKSLLE
PMPTKSIEASVANLEITEFENPWQKYMEPVDLSRNLLSTYPVELDYYYHFSVGNVCEMNEM
DFEDKEYWAKNQFHMPTGNIYGNRTPAKTNANVPALMPSKPTVRGKVKPLLILQKTVSKEHI
TPAEQGEFLAAITASSAADVFDMERLEILGDSFLKLSATLYLASKYSDWNEGTLTEVKSKLVS
NRNLLFCLIDADIPKTLNTIQFTPRYTWLPPGISLPHNVLALWRENPEFAKIIGPHNLRDLALG
DEESLVKGNCSDINYNRFVEGCRANGQSFYAGADFSSEVNFCVGLVTIPNKVIADTLEALLG
VIVKNYGLQHAFKMLEYFKICRADIDKPLTQLLNLELGGKKMRANVNTTEIDGFLINHYYLEKN
LGYTFKDRRYLLQALTHPSYPTNRITGSYQELEFIGDAILDFLISAYIFENNTKMNPGALTDLR
SALVNNTTLACICVRHRLHFFILAENAKLSEIISKFVNFQESQGHRVTNYVRILLEEADVQPTP
LDLDDELDMTELPHANKCISQEAEKGVPPKGEFNMSTNVDVPKALGDVLEALIAAVYLDCRD
==LQRTWEVIFNLFEPELQEFTRKVPINHIRQLVEHKHAKPVFSSPIVEGETVMVSCQFTCMEKT==
==IKVY==GFGSNKDQAKLSAAKHALQQLSKCDA-

Appendix 8: **Amino acid sequence of Dcr-2**
Sequences corresponding to linkers between helicase domain and DUF (yellow) as well as RNAseIII domain and dsRBD (green) are marked; cloning primers were designed for the respective DNA regions; the ΔdsRBD-Flag-Dcr-2 construct (pEH55; stock number 130) could not be detected by Western blotting, probably due to protein destabilization.

Appendix 9: Table giving an overview over stable cell culture lines

"parent cell line" indicates reporter cell line used for plasmid transfection; banmi4 = *bantam* miRNA reporter with 4 imperfect binding sites for *bantam* in the 3´UTR of GFP; bansi2 = *bantam* miRNA reporter with 2 perfect binding sites for *bantam* in the 3´UTR of GFP; 68-4 = miR-277 reporter with 4 imperfect binding sites for miR-277 in the 3´UTR of GFP; 67-1D = miR-277 reporter with 2 perfect binding sites for miR-277 in the 3´UTR of GFP; 63N1 = endo-siRNA reporter with no miRNA binding sites in the 3´UTR of GFP

transfected plasmid	parent cell line	label	transfected plasmid	parent cell line	label
pEH1	banmi4	E4	pEH2	67-1D	H3
pEH1	banmi4	F6	pEH2	67-1D	H4
pEH1	banmi4	F11	pEH2	67-1D	H5
pEH1	banmi4	F5	pEH2	67-1D	H12
pEH1	bansi2	E7	pEH2	63N1	E2
pEH1	bansi2	F3	pEH2	63N1	G9
pEH1	bansi2	F7	pEH4	banmi4	G1
pEH1	bansi2	F11	pEH4	banmi4	G6
pEH1	bansi2	H3	pEH4	banmi4	G8
pEH1	68-4	E3	pEH4	banmi4	G12
pEH1	68-4	E4	pEH4	banmi4	H1
pEH1	68-4	E8	pEH4	banmi4	H3
pEH1	68-4	E9	pEH4	banmi4	H5
pEH1	68-4	E10	pEH4	banmi4	G7
pEH1	68-4	F4	pEH4	bansi2	A2
pEH1	68-4	F8	pEH4	bansi2	A4
pEH1	68-4	F9	pEH4	bansi2	B1
pEH1	68-4	H5	pEH4	bansi2	B6
pEH1	68-4	H6	pEH4	bansi2	C11
pEH1	67-1D	G11	pEH4	bansi2	B1
pEH1	67-1D	H1	pEH4	68-4	D1
pEH1	67-1D	H11	pEH4	68-4	D4
pEH1	67-1D	H12	pEH4	63N1	E4
pEH1	67-1D	H6	pEH4	63N1	E6
pEH1	63N1	F1	pEH4	63N1	E11
pEH1	63N1	F6	pEH4	63N1	F10
pEH1	63N1	F8	pEH4	63N1	H5
pEH2	banmi4	F12	pEH7	banmi4	
pEH2	banmi4	H6	pEH7	bansi2	
pEH2	banmi4	H10	pEH7	67-1D	
pEH2	banmi4	F5	pEH7	68-4	
pEH2	banmi4	G1	pEH7	63N1	
pEH2	bansi2	F11	pEH10	banmi4	A12
pEH2	bansi2	F12	pEH10	banmi4	B1
pEH2	bansi2	G3	pEH10	banmi4	B4
pEH2	bansi2	G11	pEH10	bansi2	B2
pEH2	68-4	D10	pEH10	bansi2	B11
pEH2	68-4	E2	pEH10	68-4	A3
pEH2	68-4	F1	pEH10	68-4	B2
pEH2	67-1D	H1	pEH10	67-1D	A8

Appendix

transfected plasmid	parent cell line	label	transfected plasmid	parent cell line	label
pEH10	67-1D	B3	pEH16	67-1D	H7
pEH10	67-1D	B6	pEH16	67-1D	H8
pEH10	67-1D	B7	pEH16	63N1	A12
pEH10	63N1	B5	pEH16	63N1	A4
pEH10	63N1	B10	pEH16	63N1	B1
pEH12	banmi4	F3	pEH16	63N1	B6
pEH12	banmi4	F5	pEH16	63N1	C4
pEH12	banmi4	F6	pEH16	63N1	C8
pEH12	banmi4	F9	pEH16	63N1	D7
pEH12	banmi4	G5	pEH16	63N1	F11
pEH12	bansi2	G3	pEH18	banmi4	D2
pEH12	bansi2	G4	pEH18	banmi4	G1
pEH12	bansi2	H2	pEH18	bansi2	D2
pEH12	bansi2	H9	pEH18	bansi2	E3
pEH12	bansi2	F7	pEH18	bansi2	E5
pEH12	bansi2	G2	pEH18	bansi2	E6
pEH12	bansi2	G3	pEH18	bansi2	F8
pEH12	bansi2	G4	pEH18	bansi2	F10
pEH12	bansi2	H2	pEH18	bansi2	H8
pEH12	bansi2	H9	pEH18	68-4	E7
pEH12	bansi2	F7	pEH18	68-4	E10
pEH12	bansi2	G2	pEH18	68-4	E12
pEH12	68-4	E9	pEH18	68-4	E11
pEH12	68-4	F9	pEH18	68-4	F6
pEH12	68-4	F10	pEH18	68-4	F8
pEH12	68-4	F12	pEH18	68-4	F11
pEH12	68-4	G7	pEH18	68-4	G2
pEH12	68-4	G8	pEH18	68-4	G8
pEH12	68-4	H4	pEH18	68-4	G12
pEH12	67-1D	H4	pEH18	68-4	H11
pEH12	67-1D	H7	pEH18	67-1D	G2
pEH12	63N1	G5	pEH18	67-1D	H1
pEH12	63N1	G7	pEH18	67-1D	H8
pEH12	63N1	G8	pEH18	63N1	C11
pEH12	63N1	H2	pEH18	63N1	D4
pEH12	63N1	H1	pEH18	63N1	F1
pEH12	63N1	H4	pEH18	63N1	F4
pEH16	bansi2	D7	pEH21	banmi4	
pEH16	68-4	E6	pEH21	bansi2	
pEH16	68-4	E7	pEH21	68-4	
pEH16	68-4	F2	pEH21	67-1D	
pEH16	68-4	F5	pEH21	63N1	
pEH16	68-4	F7			
pEH16	68-4	F9			
pEH16	67-1D	G3			
pEH16	67-1D	H5			
pEH16	67-1D	H6			

9 Acknowledgements/Danksagung

Ich möchte mich herzlich bei meinem Doktorvater, Klaus Förstemann, bedanken, der mir die Möglichkeit gegeben hat, an dem Aufbau einer neuen Arbeitsgruppe beteiligt zu sein und mit dieser in meinen Erfahrungen zu wachsen. Unsere wissenschaftlichen Diskussionen haben mich sehr bereichert und begeistert, und sein profundes Wissen war ein unerschöpflicher Quell für neue Ideen. Außerdem möchte ich mich noch einmal ausdrücklich für seine Anregungen, seinen Rat und die Möglichkeiten bedanken, die er mir mit seiner Unterstützung bei Stipendien und Tagungen gegeben hat.

Eine große Bereicherung waren Prof. Dr. Gunter Meister und Prof. Dr. Michael Sattler, die mir in meinem Thesis Advisory Committee über drei Jahre hinweg mit neuen Ideen und Anregungen zur Seite standen und mich auch in Planungen über meine Doktorarbeit hinaus beraten haben.

Sehr zu Dank verpflichtet bin ich auch dem Böhringer Ingelheim Fond. Das Stipendium hat mir nicht nur ermöglicht, mehrere große internationale Konferenzen zu besuchen, sondern bot auch mit seinen internen Meetings einen ganz besonderen Rahmen. Vor allem die vielen interessanten Leute und der angeregte Austausch waren einmalig.

Ich möchte mich auch sehr herzlich bei meinen Mentorinnen des LMUMentoring Programmes, Prof. Dr. Angelika Vollmar und Prof. Dr. Christina Scheu, bedanken. Die herzlichen und sachkundigen Gespräche haben mir persönlich sehr geholfen und waren auch für meinen weiteren Karriereweg prägend. Dazu zählen auch die Möglichkeiten, interessante Fortbildungen zu besuchen und sich mit anderen Mentees auszutauschen. Ich freue mich auch sehr über einen Druckkostenzuschuss für die vorliegende Doktorarbeit durch das LMUMentoring Programm.

Mein größter Dank gilt meiner Familie, die mir immer wieder neue Kraft und Halt gegeben hat. Eure Liebe kann man nicht in Gold aufwiegen, danke, dass es euch alle gibt!

Michael Siegl, meiner studentischen Hilfskraft, möchte ich nochmal herzlich für seine gewissenhaften Fliegenkreuzungen und die nette Zusammenarbeit danken. Andreas Vogt hat als Bachelorstudent viel Mühe in Doppelstrang-RNA-Produktion gesteckt.

Außerdem möchte ich mich noch bei „meinen Mädels" in der Arbeitsgruppe und bei vielen netten Kollegen im Genzentrum bedanken. Namentlich ausdrücklich bei Joanna: morgens um 7 Uhr wird es nie wieder so lustig, traurig, spannend und anregend sein. Außerdem Diana Langer und Ursula Holter, deren Witz und Wissen, mir den besten Start in die Doktorarbeit, den man sich wünschen kann, gegeben haben.

Zu guter Letzt möchte ich auch all jenen Leuten danken, die unsere tägliche, reibungslose Arbeit in den Labors erst ermöglicht haben. Vor allem Katharina Michalik, die uns unermüdlich mit Puffern und Fliegenfutter versorgt hat.

10 References

Ahmed, F., M. Kumar and G. P. Raghava (2009). "Prediction of polyadenylation signals in human DNA sequences using nucleotide frequencies." In Silico Biol **9**(3): 135-48.

Alcorta, D. A., Y. Xiong, D. Phelps, G. Hannon, D. Beach and J. C. Barrett (1996). "Involvement of the cyclin-dependent kinase inhibitor p16 (INK4a) in replicative senescence of normal human fibroblasts." Proc Natl Acad Sci U S A **93**(24): 13742-7.

Aoki, K., H. Moriguchi, T. Yoshioka, K. Okawa and H. Tabara (2007). "In vitro analyses of the production and activity of secondary small interfering RNAs in C. elegans." Embo J **26**(24): 5007-19.

Aravin, A. A., G. J. Hannon and J. Brennecke (2007). "The Piwi-piRNA pathway provides an adaptive defense in the transposon arms race." Science **318**(5851): 761-4.

Bartel, D. P. (2004a). "MicroRNAs: genomics, biogenesis, mechanism, and function." Cell **116**(2): 281-97.

Berretta, J. and A. Morillon (2009). "Pervasive transcription constitutes a new level of eukaryotic genome regulation." EMBO Rep **10**(9): 973-82.

Blumenstiel, J. P. and D. L. Hartl (2005). "Evidence for maternally transmitted small interfering RNA in the repression of transposition in Drosophila virilis." Proc Natl Acad Sci U S A **102**(44): 15965-70.

Brand, A. H. and N. Perrimon (1993). "Targeted gene expression as a means of altering cell fates and generating dominant phenotypes." Development **118**(2): 401-15.

Brennecke, J., A. A. Aravin, A. Stark, M. Dus, M. Kellis, R. Sachidanandam and G. J. Hannon (2007). "Discrete small RNA-generating loci as master regulators of transposon activity in Drosophila." Cell **128**(6): 1089-103.

Brennecke, J., D. R. Hipfner, A. Stark, R. B. Russell and S. M. Cohen (2003). "bantam encodes a developmentally regulated microRNA that controls cell proliferation and regulates the proapoptotic gene hid in Drosophila." Cell **113**(1): 25-36.

Buszczak, M. and A. C. Spradling (2006). "The Drosophila P68 RNA helicase regulates transcriptional deactivation by promoting RNA release from chromatin." Genes Dev **20**(8): 977-89.

Chendrimada, T. P., R. I. Gregory, E. Kumaraswamy, J. Norman, N. Cooch, K. Nishikura and R. Shiekhattar (2005). "TRBP recruits the Dicer complex to Ago2 for microRNA processing and gene silencing." Nature **436**(7051): 740-4.

Chung, W. J., K. Okamura, R. Martin and E. C. Lai (2008). "Endogenous RNA interference provides a somatic defense against Drosophila transposons." Curr Biol **18**(11): 795-802.

Czech, B., C. D. Malone, R. Zhou, A. Stark, C. Schlingeheyde, M. Dus, N. Perrimon, M. Kellis, J. A. Wohlschlegel, R. Sachidanandam, G. J. Hannon and J. Brennecke (2008). "An endogenous small interfering RNA pathway in Drosophila." Nature **453**(7196): 798-802.

Daher, A., G. Laraki, M. Singh, C. E. Melendez-Pena, S. Bannwarth, A. H. Peters, E. F. Meurs, R. E. Braun, R. C. Patel and A. Gatignol (2009). "TRBP control of PACT-induced phosphorylation of protein kinase R is reversed by stress." Mol Cell Biol **29**(1): 254-65.

Daher, A., M. Longuet, D. Dorin, F. Bois, E. Segeral, S. Bannwarth, P. L. Battisti, D. F. Purcell, R. Benarous, C. Vaquero, E. F. Meurs and A. Gatignol (2001). "Two dimerization domains in the trans-activation response RNA-binding protein (TRBP) individually reverse the protein kinase R inhibition of HIV-1 long terminal repeat expression." J Biol Chem **276**(36): 33899-905.

Dalmay, T., A. Hamilton, S. Rudd, S. Angell and D. C. Baulcombe (2000). "An RNA-dependent RNA polymerase gene in Arabidopsis is required for posttranscriptional gene silencing mediated by a transgene but not by a virus." Cell **101**(5): 543-53.

Daniels, S. M., C. E. Melendez-Pena, R. J. Scarborough, A. Daher, H. S. Christensen, M. El Far, D. F. Purcell, S. Laine and A. Gatignol (2009). "Characterization of the TRBP domain required for dicer interaction and function in RNA interference." BMC Mol Biol **10**: 38.

References

Deininger, P. L. and M. A. Batzer (1999). "Alu repeats and human disease." Mol Genet Metab **67**(3): 183-93.

Di Franco, C., C. Pisano, F. Fourcade-Peronnet, G. Echalier and N. Junakovic (1992). "Evidence for de novo rearrangements of Drosophila transposable elements induced by the passage to the cell culture." Genetica **87**(2): 65-73.

Doyle, M. and M. F. Jantsch (2002). "New and old roles of the double-stranded RNA-binding domain." J Struct Biol **140**(1-3): 147-53.

Druker, R. and E. Whitelaw (2004). "Retrotransposon-derived elements in the mammalian genome: a potential source of disease." J Inherit Metab Dis **27**(3): 319-30.

Duarte, M., K. Graham, A. Daher, P. L. Battisti, S. Bannwarth, E. Segeral, K. T. Jeang and A. Gatignol (2000). "Characterization of TRBP1 and TRBP2. Stable stem-loop structure at the 5' end of TRBP2 mRNA resembles HIV-1 TAR and is not found in its processed pseudogene." J Biomed Sci **7**(6): 494-506.

Duchaine, T. F., J. A. Wohlschlegel, S. Kennedy, Y. Bei, D. Conte, Jr., K. Pang, D. R. Brownell, S. Harding, S. Mitani, G. Ruvkun, J. R. Yates, 3rd and C. C. Mello (2006). "Functional proteomics reveals the biochemical niche of C. elegans DCR-1 in multiple small-RNA-mediated pathways." Cell **124**(2): 343-54.

Finnegan, D. J., G. M. Rubin, M. W. Young and D. S. Hogness (1978). "Repeated gene families in Drosophila melanogaster." Cold Spring Harb Symp Quant Biol **42 Pt 2**: 1053-63.

Flavell, R. B., M. D. Bennett, J. B. Smith and D. B. Smith (1974). "Genome size and the proportion of repeated nucleotide sequence DNA in plants." Biochem Genet **12**(4): 257-69.

Forstemann, K., Y. Tomari, T. Du, V. V. Vagin, A. M. Denli, D. P. Bratu, C. Klattenhoff, W. E. Theurkauf and P. D. Zamore (2005). "Normal microRNA maturation and germ-line stem cell maintenance requires Loquacious, a double-stranded RNA-binding domain protein." PLoS Biol **3**(7): e236.

Förstemann, K. H., Michael D.; Wee, LiangMeng; Tomari, Yukihide and Zamore, Phillip D. (2007). "Drosophila microRNAs Are Sorted into Functionally Distinct Argonaute Complexes after Production by Dicer-1." Cell **130**(2): 287-297.

Gatignol, A., A. Buckler-White, B. Berkhout and K. T. Jeang (1991). "Characterization of a human TAR RNA-binding protein that activates the HIV-1 LTR." Science **251**(5001): 1597-600.

Ghildiyal, M., H. Seitz, M. D. Horwich, C. Li, T. Du, S. Lee, J. Xu, E. L. Kittler, M. L. Zapp, Z. Weng and P. D. Zamore (2008). "Endogenous siRNAs derived from transposons and mRNAs in Drosophila somatic cells." Science **320**(5879): 1077-81.

Ghildiyal, M. and P. D. Zamore (2009). "Small silencing RNAs: an expanding universe." Nat Rev Genet **10**(2): 94-108.

Girard, A. and G. J. Hannon (2008). "Conserved themes in small-RNA-mediated transposon control." Trends Cell Biol **18**(3): 136-48.

Girard, L. and M. Freeling (1999). "Regulatory changes as a consequence of transposon insertion." Dev Genet **25**(4): 291-6.

Golden, D. E., V. R. Gerbasi and E. J. Sontheimer (2008). "An inside job for siRNAs." Mol Cell **31**(3): 309-12.

Grishok, A., A. E. Pasquinelli, D. Conte, N. Li, S. Parrish, I. Ha, D. L. Baillie, A. Fire, G. Ruvkun and C. C. Mello (2001). "Genes and mechanisms related to RNA interference regulate expression of the small temporal RNAs that control C. elegans developmental timing." Cell **106**(1): 23-34.

Gunawardane, L. S., K. Saito, K. M. Nishida, K. Miyoshi, Y. Kawamura, T. Nagami, H. Siomi and M. C. Siomi (2007). "A slicer-mediated mechanism for repeat-associated siRNA 5' end formation in Drosophila." Science **315**(5818): 1587-90.

Gupta, V., X. Huang and R. C. Patel (2003). "The carboxy-terminal, M3 motifs of PACT and TRBP have opposite effects on PKR activity." Virology **315**(2): 283-91.

Haase, A. D., L. Jaskiewicz, H. Zhang, S. Laine, R. Sack, A. Gatignol and W. Filipowicz (2005). "TRBP, a regulator of cellular PKR and HIV-1 virus expression, interacts with Dicer and functions in RNA silencing." EMBO Rep **6**(10): 961-7.

Haley, B., G. Tang and P. D. Zamore (2003). "In vitro analysis of RNA interference in Drosophila melanogaster." Methods 30(4): 330-6.

Hall, T. A. (1999). "BioEdit: a user-friendly biological sequence alignment editor and analysis program for Windows 95/98/NT." Nucl. Acids. Symp. Ser. 41: 95-98.

Harlow, E. and D. Lane (1988). Antibodies: A Laboratory Manual. Cold Spring Harbor, Cold Spring Harbor Laboratory.

Hartig, J. V., S. Esslinger, R. Bottcher, K. Saito and K. Forstemann (2009). "Endo-siRNAs depend on a new isoform of loquacious and target artificially introduced, high-copy sequences." EMBO J 28(19): 2932-2944.

Hartig, J. V. and K. Förstemann (submitted). "Loqs-PD and R2D2 commit Dcr-2 to the endo- or exo-siRNA pathway in Drosophila."

Hartig, J. V., Y. Tomari and K. Forstemann (2007). "piRNAs--the ancient hunters of genome invaders." Genes Dev 21(14): 1707-13.

Huang, X., B. Hutchins and R. C. Patel (2002). "The C-terminal, third conserved motif of the protein activator PACT plays an essential role in the activation of double-stranded-RNA-dependent protein kinase (PKR)." Biochem J 366(Pt 1): 175-86.

Hutvagner, G., J. McLachlan, A. E. Pasquinelli, E. Balint, T. Tuschl and P. D. Zamore (2001). "A cellular function for the RNA-interference enzyme Dicer in the maturation of the let-7 small temporal RNA." Science 293(5531): 834-8.

Ishizuka, A., M. C. Siomi and H. Siomi (2002). "A Drosophila fragile X protein interacts with components of RNAi and ribosomal proteins." Genes Dev 16(19): 2497-508.

Jiang, F., X. Ye, X. Liu, L. Fincher, D. McKearin and Q. Liu (2005). "Dicer-1 and R3D1-L catalyze microRNA maturation in Drosophila." Genes Dev 19(14): 1674-9.

Kapranov, P., J. Cheng, S. Dike, D. A. Nix, R. Duttagupta, A. T. Willingham, P. F. Stadler, J. Hertel, J. Hackermuller, I. L. Hofacker, I. Bell, E. Cheung, J. Drenkow, E. Dumais, S. Patel, G. Helt, M. Ganesh, S. Ghosh, A. Piccolboni, V. Sementchenko, H. Tammana and T. R. Gingeras (2007). "RNA maps reveal new RNA classes and a possible function for pervasive transcription." Science 316(5830): 1484-8.

Kawamura, Y., K. Saito, T. Kin, Y. Ono, K. Asai, T. Sunohara, T. N. Okada, M. C. Siomi and H. Siomi (2008). "Drosophila endogenous small RNAs bind to Argonaute 2 in somatic cells." Nature 453(7196): 793-7.

Kazazian, H. H., Jr. (1999). "An estimated frequency of endogenous insertional mutations in humans." Nat Genet 22(2): 130.

Kazazian, H. H., Jr. (2004). "Mobile elements: drivers of genome evolution." Science 303(5664): 1626-32.

Kennedy, S., D. Wang and G. Ruvkun (2004). "A conserved siRNA-degrading RNase negatively regulates RNA interference in C. elegans." Nature 427(6975): 645-9.

Ketting, R. F., S. E. Fischer, E. Bernstein, T. Sijen, G. J. Hannon and R. H. Plasterk (2001). "Dicer functions in RNA interference and in synthesis of small RNA involved in developmental timing in C. elegans." Genes Dev 15(20): 2654-9.

Kim, J. M., S. Vanguri, J. D. Boeke, A. Gabriel and D. F. Voytas (1998). "Transposable elements and genome organization: a comprehensive survey of retrotransposons revealed by the complete Saccharomyces cerevisiae genome sequence." Genome Res 8(5): 464-78.

Kim, V. N., J. Han and M. C. Siomi (2009). "Biogenesis of small RNAs in animals." Nat Rev Mol Cell Biol 10(2): 126-39.

Knight, S. W. and B. L. Bass (2001). "A role for the RNase III enzyme DCR-1 in RNA interference and germ line development in Caenorhabditis elegans." Science 293(5538): 2269-71.

Kooter, J. M., M. A. Matzke and P. Meyer (1999). "Listening to the silent genes: transgene silencing, gene regulation and pathogen control." Trends Plant Sci 4(9): 340-347.

Lander, E. S., L. M. Linton, B. Birren, C. Nusbaum, M. C. Zody, J. Baldwin, K. Devon, K. Dewar, M. Doyle, W. FitzHugh, R. Funke, D. Gage, K. Harris, A. Heaford, J. Howland, L. Kann, J. Lehoczky, R. LeVine, P. McEwan, K. McKernan, J. Meldrim, J. P. Mesirov, C. Miranda, W. Morris, J.

Naylor, C. Raymond, M. Rosetti, R. Santos, A. Sheridan, C. Sougnez, N. Stange-Thomann, N. Stojanovic, A. Subramanian, D. Wyman, J. Rogers, J. Sulston, R. Ainscough, S. Beck, D. Bentley, J. Burton, C. Clee, N. Carter, A. Coulson, R. Deadman, P. Deloukas, A. Dunham, I. Dunham, R. Durbin, L. French, D. Grafham, S. Gregory, T. Hubbard, S. Humphray, A. Hunt, M. Jones, C. Lloyd, A. McMurray, L. Matthews, S. Mercer, S. Milne, J. C. Mullikin, A. Mungall, R. Plumb, M. Ross, R. Shownkeen, S. Sims, R. H. Waterston, R. K. Wilson, L. W. Hillier, J. D. McPherson, M. A. Marra, E. R. Mardis, L. A. Fulton, A. T. Chinwalla, K. H. Pepin, W. R. Gish, S. L. Chissoe, M. C. Wendl, K. D. Delehaunty, T. L. Miner, A. Delehaunty, J. B. Kramer, L. L. Cook, R. S. Fulton, D. L. Johnson, P. J. Minx, S. W. Clifton, T. Hawkins, E. Branscomb, P. Predki, P. Richardson, S. Wenning, T. Slezak, N. Doggett, J. F. Cheng, A. Olsen, S. Lucas, C. Elkin, E. Uberbacher, M. Frazier, R. A. Gibbs, D. M. Muzny, S. E. Scherer, J. B. Bouck, E. J. Sodergren, K. C. Worley, C. M. Rives, J. H. Gorrell, M. L. Metzker, S. L. Naylor, R. S. Kucherlapati, D. L. Nelson, G. M. Weinstock, Y. Sakaki, A. Fujiyama, M. Hattori, T. Yada, A. Toyoda, T. Itoh, C. Kawagoe, H. Watanabe, Y. Totoki, T. Taylor, J. Weissenbach, R. Heilig, W. Saurin, F. Artiguenave, P. Brottier, T. Bruls, E. Pelletier, C. Robert, P. Wincker, D. R. Smith, L. Doucette-Stamm, M. Rubenfield, K. Weinstock, H. M. Lee, J. Dubois, A. Rosenthal, M. Platzer, G. Nyakatura, S. Taudien, A. Rump, H. Yang, J. Yu, J. Wang, G. Huang, J. Gu, L. Hood, L. Rowen, A. Madan, S. Qin, R. W. Davis, N. A. Federspiel, A. P. Abola, M. J. Proctor, R. M. Myers, J. Schmutz, M. Dickson, J. Grimwood, D. R. Cox, M. V. Olson, R. Kaul, C. Raymond, N. Shimizu, K. Kawasaki, S. Minoshima, G. A. Evans, M. Athanasiou, R. Schultz, B. A. Roe, F. Chen, H. Pan, J. Ramser, H. Lehrach, R. Reinhardt, W. R. McCombie, M. de la Bastide, N. Dedhia, H. Blocker, K. Hornischer, G. Nordsiek, R. Agarwala, L. Aravind, J. A. Bailey, A. Bateman, S. Batzoglou, E. Birney, P. Bork, D. G. Brown, C. B. Burge, L. Cerutti, H. C. Chen, D. Church, M. Clamp, R. R. Copley, T. Doerks, S. R. Eddy, E. E. Eichler, T. S. Furey, J. Galagan, J. G. Gilbert, C. Harmon, Y. Hayashizaki, D. Haussler, H. Hermjakob, H. Hokamp, W. Jang, L. S. Johnson, T. A. Jones, S. Kasif, A. Kaspryzk, S. Kennedy, W. J. Kent, P. Kitts, E. V. Koonin, I. Korf, D. Kulp, D. Lancet, T. M. Lowe, A. McLysaght, T. Mikkelsen, J. V. Moran, N. Mulder, V. J. Pollara, C. P. Ponting, G. Schuler, J. Schultz, G. Slater, A. F. Smit, E. Stupka, J. Szustakowski, D. Thierry-Mieg, J. Thierry-Mieg, L. Wagner, J. Wallis, R. Wheeler, A. Williams, Y. I. Wolf, K. H. Wolfe, S. P. Yang, R. F. Yeh, F. Collins, M. S. Guyer, J. Peterson, A. Felsenfeld, K. A. Wetterstrand, A. Patrinos, M. J. Morgan, P. de Jong, J. J. Catanese, K. Osoegawa, H. Shizuya, S. Choi and Y. J. Chen (2001). "Initial sequencing and analysis of the human genome." Nature **409**(6822): 860-921.

Laraki, G., G. Clerzius, A. Daher, C. Melendez-Pena, S. Daniels and A. Gatignol (2008). "Interactions between the double-stranded RNA-binding proteins TRBP and PACT define the Medipal domain that mediates protein-protein interactions." RNA Biol **5**(2): 92-103.

Lau, P. W., C. S. Potter, B. Carragher and I. J. MacRae (2009). "Structure of the human Dicer-TRBP complex by electron microscopy." Structure **17**(10): 1326-32.

Le, T., M. Yu, B. Williams, S. Goel, S. M. Paul and G. J. Beitel (2007). "CaSpeR5, a family of Drosophila transgenesis and shuttle vectors with improved multiple cloning sites." Biotechniques **42**(2): 164, 166.

Lee, R. C., R. L. Feinbaum and V. Ambros (1993). "The C. elegans heterochronic gene lin-4 encodes small RNAs with antisense complementarity to lin-14." Cell **75**(5): 843-54.

Lee, Y., I. Hur, S. Y. Park, Y. K. Kim, M. R. Suh and V. N. Kim (2006). "The role of PACT in the RNA silencing pathway." Embo J **25**(3): 522-32.

Lee, Y., K. Jeon, J. T. Lee, S. Kim and V. N. Kim (2002). "MicroRNA maturation: stepwise processing and subcellular localization." Embo J **21**(17): 4663-70.

Lee, Y., M. Kim, J. Han, K. H. Yeom, S. Lee, S. H. Baek and V. N. Kim (2004). "MicroRNA genes are transcribed by RNA polymerase II." Embo J **23**(20): 4051-60.

Lee, Y. S., K. Nakahara, J. W. Pham, K. Kim, Z. He, E. J. Sontheimer and R. W. Carthew (2004). "Distinct roles for Drosophila Dicer-1 and Dicer-2 in the siRNA/miRNA silencing pathways." Cell **117**(1): 69-81.

Lim do, H., J. Kim, S. Kim, R. W. Carthew and Y. S. Lee (2008). "Functional analysis of dicer-2 missense mutations in the siRNA pathway of Drosophila." Biochem Biophys Res Commun **371**(3): 525-30.

Lipardi, C. and B. M. Paterson (2009). "Identification of an RNA-dependent RNA polymerase in Drosophila involved in RNAi and transposon suppression." Proc Natl Acad Sci U S A **106**(37): 15645-50.

Lipardi, C., Q. Wei and B. M. Paterson (2001). "RNAi as random degradative PCR: siRNA primers convert mRNA into dsRNAs that are degraded to generate new siRNAs." Cell **107**(3): 297-307.

Liu, Q., T. A. Rand, S. Kalidas, F. Du, H. E. Kim, D. P. Smith and X. Wang (2003). "R2D2, a bridge between the initiation and effector steps of the Drosophila RNAi pathway." Science **301**(5641): 1921-5.

Liu, X., F. Jiang, S. Kalidas, D. Smith and Q. Liu (2006). "Dicer-2 and R2D2 coordinately bind siRNA to promote assembly of the siRISC complexes." Rna **12**(8): 1514-20.

Liu, X., J. K. Park, F. Jiang, Y. Liu, D. McKearin and Q. Liu (2007). "Dicer-1, but not Loquacious, is critical for assembly of miRNA-induced silencing complexes." Rna **13**(12): 2324-9.

Lodish, H., A. Berk, S. L. Zipursky, P. Matsudaira, D. Baltimore and J. Darnell (2000). Molecular Cell Biology (4th edition). New York, NY.

Ma, E., I. J. MacRae, J. F. Kirsch and J. A. Doudna (2008). "Autoinhibition of human dicer by its internal helicase domain." J Mol Biol **380**(1): 237-43.

Maisonhaute, C., D. Ogereau, A. Hua-Van and P. Capy (2007). "Amplification of the 1731 LTR retrotransposon in Drosophila melanogaster cultured cells: origin of neocopies and impact on the genome." Gene **393**(1-2): 116-26.

Malone, C. D. and G. J. Hannon (2009). "Small RNAs as guardians of the genome." Cell **136**(4): 656-68.

Marques, J. T., K. Kim, P. H. Wu, T. M. Alleyne, N. Jafari and R. W. Carthew (2010). "Loqs and R2D2 act sequentially in the siRNA pathway in Drosophila." Nat Struct Mol Biol **17**(1): 24-30.

Martinez, J., A. Patkaniowska, H. Urlaub, R. Luhrmann and T. Tuschl (2002). "Single-stranded antisense siRNAs guide target RNA cleavage in RNAi." Cell **110**(5): 563-74.

Mette, M. F., W. Aufsatz, J. van der Winden, M. A. Matzke and A. J. Matzke (2000). "Transcriptional silencing and promoter methylation triggered by double-stranded RNA." Embo J **19**(19): 5194-201.

Miyoshi, K., T. Miyoshi, J. V. Hartig, H. Siomi and M. C. Siomi (2010). "Molecular mechanisms that funnel RNA precursors into endogenous small-interfering RNA and microRNA biogenesis pathways in Drosophila." Rna.

Miyoshi, K., T. N. Okada, H. Siomi and M. C. Siomi (2009). "Characterization of the miRNA-RISC loading complex and miRNA-RISC formed in the Drosophila miRNA pathway." Rna **15**(7): 1282-91.

Nakamura, M., R. Ando, T. Nakazawa, T. Yudazono, N. Tsutsumi, N. Hatanaka, T. Ohgake, F. Hanaoka and T. Eki (2007). "Dicer-related drh-3 gene functions in germ-line development by maintenance of chromosomal integrity in Caenorhabditis elegans." Genes Cells **12**(9): 997-1010.

Nykanen, A., B. Haley and P. D. Zamore (2001). "ATP requirements and small interfering RNA structure in the RNA interference pathway." Cell **107**(3): 309-21.

Okamura, K., S. Balla, R. Martin, N. Liu and E. C. Lai (2008a). "Two distinct mechanisms generate endogenous siRNAs from bidirectional transcription in Drosophila melanogaster." Nat Struct Mol Biol **15**(6): 581-90.

Okamura, K., W. J. Chung, J. G. Ruby, H. Guo, D. P. Bartel and E. C. Lai (2008b). "The Drosophila hairpin RNA pathway generates endogenous short interfering RNAs." Nature **453**(7196): 803-6.

Okamura, K., A. Ishizuka, H. Siomi and M. C. Siomi (2004). "Distinct roles for Argonaute proteins in small RNA-directed RNA cleavage pathways." Genes Dev **18**(14): 1655-66.

Okamura, K. and E. C. Lai (2008c). "Endogenous small interfering RNAs in animals." Nat Rev Mol Cell Biol **9**(9): 673-8.

Pak, J. and A. Fire (2007). "Distinct populations of primary and secondary effectors during RNAi in C. elegans." Science 315(5809): 241-4.

Park, H., M. V. Davies, J. O. Langland, H. W. Chang, Y. S. Nam, J. Tartaglia, E. Paoletti, B. L. Jacobs, R. J. Kaufman and S. Venkatesan (1994). "TAR RNA-binding protein is an inhibitor of the interferon-induced protein kinase PKR." Proc Natl Acad Sci U S A 91(11): 4713-7.

Park, J. K., X. Liu, T. J. Strauss, D. M. McKearin and Q. Liu (2007). "The miRNA pathway intrinsically controls self-renewal of Drosophila germline stem cells." Curr Biol 17(6): 533-8.

Parker, G. S., D. M. Eckert and B. L. Bass (2006). "RDE-4 preferentially binds long dsRNA and its dimerization is necessary for cleavage of dsRNA to siRNA." Rna 12(5): 807-18.

Parker, G. S., T. S. Maity and B. L. Bass (2008). "dsRNA binding properties of RDE-4 and TRBP reflect their distinct roles in RNAi." J Mol Biol 384(4): 967-79.

Patel, R. C. and G. C. Sen (1998). "PACT, a protein activator of the interferon-induced protein kinase, PKR." Embo J 17(15): 4379-90.

Pham, J. W., J. L. Pellino, Y. S. Lee, R. W. Carthew and E. J. Sontheimer (2004). "A Dicer-2-dependent 80s complex cleaves targeted mRNAs during RNAi in Drosophila." Cell 117(1): 83-94.

Potter, S. S., W. J. Brorein, Jr., P. Dunsmuir and G. M. Rubin (1979). "Transposition of elements of the 412, copia and 297 dispersed repeated gene families in Drosophila." Cell 17(2): 415-27.

Retelska, D., C. Iseli, P. Bucher, C. V. Jongeneel and F. Naef (2006). "Similarities and differences of polyadenylation signals in human and fly." BMC Genomics 7: 176.

Roignant, J. Y., C. Carre, B. Mugat, D. Szymczak, J. A. Lepesant and C. Antoniewski (2003). "Absence of transitive and systemic pathways allows cell-specific and isoform-specific RNAi in Drosophila." Rna 9(3): 299-308.

Ryter, J. M. and S. C. Schultz (1998). "Molecular basis of double-stranded RNA-protein interactions: structure of a dsRNA-binding domain complexed with dsRNA." Embo J 17(24): 7505-13.

Saito, K., A. Ishizuka, H. Siomi and M. C. Siomi (2005). "Processing of pre-microRNAs by the Dicer-1-Loquacious complex in Drosophila cells." PLoS Biol 3(7): e235.

Schmittgen, T. D. and K. J. Livak (2008). "Analyzing real-time PCR data by the comparative C(T) method." Nat Protoc 3(6): 1101-8.

Schneider, I. (1972). "Cell lines derived from late embryonic stages of Drosophila melanogaster." J Embryol Exp Morphol 27(2): 353-65.

Seitz, H., M. Ghildiyal and P. D. Zamore (2008). "Argonaute loading improves the 5' precision of both MicroRNAs and their miRNA strands in flies." Curr Biol 18(2): 147-51.

Shah, C. and K. Forstemann (2008). "Monitoring miRNA-mediated silencing in Drosophila melanogaster S2-cells." Biochim Biophys Acta 1779(11): 766-72.

Shalgi, R., M. Lapidot, R. Shamir and Y. Pilpel (2005). "A catalog of stability-associated sequence elements in 3' UTRs of yeast mRNAs." Genome Biol 6(10): R86.

Sijen, T., J. Fleenor, F. Simmer, K. L. Thijssen, S. Parrish, L. Timmons, R. H. Plasterk and A. Fire (2001). "On the role of RNA amplification in dsRNA-triggered gene silencing." Cell 107(4): 465-76.

Sijen, T., F. A. Steiner, K. L. Thijssen and R. H. Plasterk (2007). "Secondary siRNAs result from unprimed RNA synthesis and form a distinct class." Science 315(5809): 244-7.

Simmer, F., M. Tijsterman, S. Parrish, S. P. Koushika, M. L. Nonet, A. Fire, J. Ahringer and R. H. Plasterk (2002). "Loss of the putative RNA-directed RNA polymerase RRF-3 makes C. elegans hypersensitive to RNAi." Curr Biol 12(15): 1317-9.

Stefl, R., M. Xu, L. Skrisovska, R. B. Emeson and F. H. Allain (2006). "Structure and specific RNA binding of ADAR2 double-stranded RNA binding motifs." Structure 14(2): 345-55.

Tabara, H., E. Yigit, H. Siomi and C. C. Mello (2002). "The dsRNA binding protein RDE-4 interacts with RDE-1, DCR-1, and a DExH-box helicase to direct RNAi in C. elegans." Cell 109(7): 861-71.

Tomari, Y., T. Du, B. Haley, D. S. Schwarz, R. Bennett, H. A. Cook, B. S. Koppetsch, W. E. Theurkauf and P. D. Zamore (2004). "RISC assembly defects in the Drosophila RNAi mutant armitage." Cell 116(6): 831-41.

Tomari, Y. D., Tingting and Zamore, Phillip D. (2007). "Sorting of Drosophila Small Silencing RNAs." Cell 130(2): 299-308.

Tuschl, T., P. D. Zamore, R. Lehmann, D. P. Bartel and P. A. Sharp (1999). "Targeted mRNA degradation by double-stranded RNA in vitro." Genes Dev **13**(24): 3191-7.

Wang, H. W., C. Noland, B. Siridechadilok, D. W. Taylor, E. Ma, K. Felderer, J. A. Doudna and E. Nogales (2009). "Structural insights into RNA processing by the human RISC-loading complex." Nat Struct Mol Biol **16**(11): 1148-53.

Xie, X., J. Lu, E. J. Kulbokas, T. R. Golub, V. Mootha, K. Lindblad-Toh, E. S. Lander and M. Kellis (2005). "Systematic discovery of regulatory motifs in human promoters and 3' UTRs by comparison of several mammals." Nature **434**(7031): 338-45.

Ye, X., Z. Paroo and Q. Liu (2007). "Functional anatomy of the Drosophila microRNA-generating enzyme." J Biol Chem **282**(39): 28373-8.

Yigit, E., P. J. Batista, Y. Bei, K. M. Pang, C. C. Chen, N. H. Tolia, L. Joshua-Tor, S. Mitani, M. J. Simard and C. C. Mello (2006). "Analysis of the C. elegans Argonaute family reveals that distinct Argonautes act sequentially during RNAi." Cell **127**(4): 747-57.

Zamore, P. D., T. Tuschl, P. A. Sharp and D. P. Bartel (2000). "RNAi: double-stranded RNA directs the ATP-dependent cleavage of mRNA at 21 to 23 nucleotide intervals." Cell **101**(1): 25-33.

Zhou, R., B. Czech, J. Brennecke, R. Sachidanandam, J. A. Wohlschlegel, N. Perrimon and G. J. Hannon (2009). "Processing of Drosophila endo-siRNAs depends on a specific Loquacious isoform." Rna **15**(10): 1886-95.

Die VDM Verlagsservicegesellschaft sucht für wissenschaftliche Verlage abgeschlossene und herausragende

Dissertationen, Habilitationen, Diplomarbeiten, Master Theses, Magisterarbeiten usw.

für die kostenlose Publikation als Fachbuch.

Sie verfügen über eine Arbeit, die hohen inhaltlichen und formalen Ansprüchen genügt, und haben Interesse an einer honorarvergüteten Publikation?

Dann senden Sie bitte erste Informationen über sich und Ihre Arbeit per Email an *info@vdm-vsg.de*.

Sie erhalten kurzfristig unser Feedback!

VDM Verlagsservicegesellschaft mbH
Dudweiler Landstr. 99   Telefon +49 681 3720 174
D - 66123 Saarbrücken   Fax   +49 681 3720 1749
www.vdm-vsg.de

Die VDM Verlagsservicegesellschaft mbH vertritt

Printed by Books on Demand GmbH, Norderstedt / Germany